On Scientific Representations

Also by Giovanni Boniolo

MATHEMATICS AND PHYSICS (2005)

EVOLUTIONARY ETHICS AND CONTEMPORARY BIOLOGY (2006)

On Scientific Representations

From Kant to a New Philosophy of Science

Giovanni Boniolo

palgrave
macmillan

First published 2007 by
PALGRAVE MACMILLAN
Houndmills, Basingstoke, Hampshire RG21 6XS and
175 Fifth Avenue, New York, N.Y. 10010
Companies and representatives throughout the world

PALGRAVE MACMILLAN is the global academic imprint of the Palgrave Macmillan division of St. Martin's Press, LLC and of Palgrave Macmillan Ltd. Macmillan® is a registered trademark in the United States, United Kingdom and other countries. Palgrave is a registered trademark in the European Union and other countries.

ISBN-13: 978–0–230–52208–4 hardback
ISBN-10: 0–230–52208–4 hardback

This book is printed on paper suitable for recycling and made from fully managed and sustained forest sources. Logging, pulping and manufacturing processes are expected to conform to the environmental regulations of the country of origin

A catalogue record for this book is available from the British Library.

Library of Congress Cataloging-in-Publication Data
Boniolo, Giovanni.
On scientific representations:from Kant to a new philosophy of science/Giovanni Boniolo.
p. cm.
Includes bibliographical references and index.
ISBN-10: 0–230–52208–4 (cloth)
ISBN-13: 978–0–230–52208–4 (cloth)
1. Science—Philosophy. 2. Knowledge, Theory of. I. Title.
Q175.B7223 2007
501—dc22 2006052512

10 9 8 7 6 5 4 3 2 1
16 15 14 13 12 11 10 09 08 07

Printed and bound in Great Britain by
Antony Rowe Ltd, Chippenham and Eastbourne

To Giorgia, Fabio, Irene

Contents

List of Figures

Acknowledgements

I should mention all those friends, students, colleagues and unknown curious people who have read parts of this book or attended lectures, meetings and conferences in which I presented some positions here contained, and who argued with them or asked for clarifications. There are too many persons to recall, and to thank one by one. I could acknowledge only the most renowned ones, but it would be unjust and unfair to those others who sometimes made better objections and posed better questions. I want to say a general 'Thanks to all of you'.

However, let me express my enormous pleasure in having met very brilliant contemporary thinkers, from whom I learned and with whom I discussed a lot, and my happiness and delight in having read illuminating pages of books and papers.

Nevertheless, only one name: Paola, my wife.

GIOVANNI BONIOLO

Introduction

Conoscimus
non ipsam speciem impressam,
sed per speciem.

General

The increasing power of the Nazis in Europe before the beginning of World War II had serious negative consequences on the international philosophical community, in particular on philosophers of science. Before that time, the philosophical community was polyglot and the main official languages were English, French, German and Italian; international philosophical journals accepted papers in all these languages without any problem and the philosophical atmosphere was intellectually and culturally extremely various. Each school of thought and each individual scholar tried to construct their own way of doing philosophy in their own style, but it was always deeply rooted in the history of philosophy. Authors such as Ardigò and Enriques in Italy, Poincaré and Duhem in France, Eddington and Campbell in Great Britain, Cassirer, Hertz, Mach, Boltzmann, and (at least in the early years of their work) the neopositivists in the German-speaking countries, did not deny their place as descendants of a very ancient philosophical tradition. They did not ignore the great philosophers of the past, even if they were extremely creative in proposing new philosophical horizons.

Because of the rise of Nazism, war, and the consequent flow of many neopositivists into the English-speaking countries, all of this came to an end. Almost immediately the official language of philosophy became English and soon afterwards the historical roots of philosophy were forgotten, or simply neglected. This was also due to the (meta-philosophical) presupposition that philosophy should be thought of as an activity aimed at explicating through logico-linguistic tools otherwise blurry and obscure concepts; an activity which would require neither profound historical foundations nor historical awareness. Moreover, the shift of the geographical core of the production of philosophy from continental Europe to the English-speaking countries led to a shift in the intellectual background knowledge: from many classical traditions (Aristotelian, Cartesian, Humean, Kantian, the Phenomenological tradition, and so on) to a very few (the Humean, and in some cases, the pre-Humean).

Now things seem to be changing. Even if the official language of philosophy is still English, philosophers have become more attentive to their great philosophical tradition: there are, for example, a Kantian renaissance and a Phenomenological renaissance. Moreover contemporary philosophers appear to be less 'ashamed' of their historical roots. It seems that we philosophers again feel the need to be a part of an ancient tradition of thought.

The overarching aims of *On Scientific Representations* are placed squarely within this cultural framework. In particular, I hope to demonstrate that it is possible to produce new philosophical proposals without forgetting the historical heritage. I strongly believe that doing philosophy means (1) having problems historically rooted, (2) proposing solutions that are historically aware, and (3) arguing cogently for those solutions. This means that I support the idea that each real philosophical problem has a history, and that it is strictly connected with mankind having needs, passions, desires, and above all traditions. Philosophical problems need a historical context to give them life.

Certainly, the originality of the solutions proposed could be hidden by their history, but this is a risk we should run. Coherent with this approach, from the history of the past I try to propose new ideas. These could be dismissed as spin-offs; but I think that everything we can propose as a novelty is actually a spin-off of something already proposed. It could not be otherwise. It is possible to write a book on a philosophical subject without quoting anyone from the past but only through ignorance of the history of philosophy, the history of the intellectual past or of tradition (and sometimes this happens in the contemporary philosophical scenario), or because a writer prefers to act as though there were no tradition in philosophy beyond his or her own (and sometimes this also happens). I support a more modest historical style. I prefer to root my view in history because I am proud of the tradition of thought to which I belong, and because I think that the best way (and maybe the only way) of maintaining our own cultural identity, especially today, consists in not losing the link with our past. We are the sons and daughters of our history. A philosopher cannot be human without history, a barbarian who has just learned the language. On the contrary, a philosopher is as old as the philosophical tradition. Our philosopher was born around 3000 years ago in the Middle East and grew up in Greece and in Rome. Adolescence was spent studying in medieval Oxford, Paris and Padova. Then he was a 'cleric' who visited England, France, Germany,

Ireland, Italy, Scotland and so on. Now the philosopher is likely to live in the English-speaking countries. Nobody denies this latter fact, but neither must we deny its history.

This is why, in *On Scientific Representations* there is such emphasis on the historical reconstruction of the questions tackled. It is by returning to our intellectual ancestors that we may receive new lifeblood for our efforts philosophically to comprehend the world we live in and the world we come from. In particular, as will become clear, the philosophical ancestor to whom I suggest we return is Kant. More precisely, I suggest reading Kant and the Kantians again, and regarding them as the starting point for rethinking and renewing some contemporary issues in philosophy, especially in epistemology and the philosophy of science.

On Scientific Representations then, is an eccentric philosophical work, precisely in the etymological sense of eccentric, that is, far from the centre. First, it is eccentric because it proposes a new way of considering some philosophical questions – by starting from history – and this is not exactly what the philosophical fashion imposes nowadays, even if a new importance is beginning to be given to the history of philosophy, that is, to our tradition. It is eccentric in a second sense, since it conjugates the historical tradition with the theoretical tradition. Perhaps I should have used the terms 'continental' and 'analytical', but, to be honest, I am not entirely sure of the real meaning of these two locutions. However, it is undeniable that many parts of *On Scientific Representations* are written as though by a historically-minded philosopher; but it is undeniable as well, that many parts of *On Scientific Representations* are written as though by a theoretically-minded philosopher. It is eccentric also in a third sense. Contemporary philosophy seems to be mostly Humean-like, or a pre-Humean-like philosophy. Instead *On Scientific Representations* presents a Kantian point of view. Contemporary philosophy is mostly made in an ahistorical furrow by Humean or pre-Humean ploughmen. My attempt consists simply in moving the plough from that furrow and in trying to begin another one. I will insert the issues of scientific concepts, laws, theories, models and thought experiments into a unique coherent framework with a strong Kantian flavour. Starting from a precise historical reconstruction of the Kantian tradition on scientific representations (Kant, Hertz, Vaihinger, Cassirer and so on), I offer an original and promising approach to the philosophy of science. And last, the book is eccentric in a fourth sense. As far as I know, there have been few attempts to realize a happy marriage between epistemology and the philosophy of science (and science). Often, contemporary philosophers of science do not consider the epistemological side of their work, and, conversely, many contemporary epistemologists do not consider the philosophy of science. I propose, instead, a unitary approach, and position myself in a tradition characterized by this union: there cannot be epistemology without a philosophy of science (and science), and there cannot be a philosophy of science without epistemology.

Plan of the book

Core idea

The main goal of *On Scientific Representations* concerns the analysis of the notion of *representation* from a Kantian approach. This is developed in three chapters, each dealing with a different aspect of the way of representing the world, that is, (1) scientific concepts as representations, and objects as something cognitively due to a synthesis of representations, (2) scientific laws as representations, (3) scientific theories and scientific models as representations.

Chapter 1: Concepts and objects

'What is a concept?': hard and *vexata quaestio*, although if we look at the history of philosophy, we would find many possible answers. By following a leading thread, I will focus only on some of them, and then I will propose my own, connected to the same leading thread. Such a leading thread is given by Cassirer's effort to fuse together Kant's doctrine of concept with Frege's. In so doing, especially in *Substance and Function*, he offers a very stringent critique of Aristotle's classical theory of concept. By discussing Cassirer's works, it will be clear that there is a wide gap between the classical theory, which considers *concepts as classes* (class-concept), and some of the modern and contemporary theories, which consider *concepts as functions* (function-concepts).

In the first part of the chapter, I will present the *status quaestionis*. This means that I will discuss from a historical-critical viewpoint the main features of four proposals: Aristotle's, Kant's, Frege's and Cassirer's. The discussion aims both to show their weaknesses and to disentangle possible terminological ambiguities (in particular, 'function' à la Kant and à la Frege); through this discussion I explicitly state the tradition into which my proposal fits (a Kantian tradition). In the second part of the chapter I will present and argue my thesis, according to which *concepts are both representations and rules*.

Although in this second part the core aim is to tackle the analysis of the notion of 'concept', it will also be connected to the discussion of the notion of 'object', which will lead to the idea of *objects as knots of properties*. Underlying this is always a Kantian perspective, an emphasis on the cognitive constitution of meaning of the world due to the imposition of the knowing subject's conceptual apparatus. I will demonstrate that this approach is extremely relevant, with reference to the ontology of contemporary scientific theories.

Finally, and on the bases of the previous analyses, I will propose a particular form of realism, which I call *wise realism*, that is able to escape both from the Kantian dualism between noumenon and phenomenon, and from naive forms of (on the one hand) constructivism or (on the other hand) mirror-like theses.

Chapter 2: Laws of nature

If we reflect on the birth of the contemporary debate concerning laws of nature, we realize that it was the neopositivist movement which shifted it from the epistemological role of the laws to the problem of the logical and epistemological characterizations of nomological statements. In particular it is M. Schlick, with his 1931 'Causality in Contemporary Physics', whom I consider the 'father' of this debate. Starting from an analysis of Schlick's paper, I will recall the failure of the neopositivist and post-positivist regularists to demarcate an accidental generalization from a nomological generalization. After reviewing this Humean approach, I will move to the contemporary metaphysical attempts to solve the problem, and to their failure. It should be noted that if the regularists can be considered Humean, the metaphysicians can be considered pre-Humean, in part because they ground nomologicity into something rather obscure like metaphysical necessity.

It is evident, of course, for those who know the history of their tradition, that this contemporary debate has forgotten the Kantian approach. From this point of view, laws of nature are no longer a mere regularity or, even worse, something metaphysical, but something that the knowing subject imposes on the world to constitute it (or part of it) as a cognitively ordinate set of phenomena. To present this point of view I will recall what Kant said in his critical works. Moreover, by reconstructing Kant's position, I will show the usefulness of a Kantian way of thinking in solving the problem of the laws of nature, and how innovative it is with respect both to the regularist and to the metaphysical approach. In particular I will develop a Kantian solution to the problem of laws by emphasizing the role of the knowing subject and the notion of *system of statements*. In this way, also the concepts of *lawness* (concerning the conditions according to which a lawlike statement is to be considered as a law) and *lawfulness* (regarding what is allowed and not allowed in nature, intended à la Kant) will be discussed and explicated.

Chapter 3: Theories, models, thought experiments, and counterfactuals

This chapter is divided in three parts. In the first part the issue of the relations between scientific theories and scientific models is considered. This might be considered as old hat; something on which nothing new can be said, but this is not true. I will propose an interpretation of this issue according to which *scientific theories are hypothetical representations*, that is, conceptual tools made for cognitive aims, and *models are fictions*, or, rather, *fictive representations*, that is, 'as-if' constructions, or conceptual tools devised for pragmatic aims. This means that the notion of 'fiction' will be extensively analysed and correctly housed in the underestimated Vaihinger's *The Philosophy of 'as if'*.

In the second part of this chapter, the topic of thought experiments will be discussed and grounded on the notion of 'as-if world'. The two classes into which they can be differentiated will be shown: the rhetorical thought experiments and the exploratory/clarifying thought experiments. The thought experiments of the first class are used for rhetorical aims. They are characterized by an argument starting from an 'as-if-world' and by means of which a thesis is supported or opposed. The thought experiments of the second class are constructed for theoretical aims, being intellectual tools by means of which certain theoretical questions are explored or clarified by reflecting on 'as-if-worlds'.

The third part of the chapter will be devoted to correlating the notion of 'fiction' to the notion of 'counterfactual' and to offering a new approach to the topic. In particular the so-called counterfactual problem, that is, the problem concerning the validity of a counterfactual proposition, will be approached not in the usual logico-linguistic way, but in a totally original way, that is, from a *hermeneutical approach*. It will be argued how innovative and powerful this different analytical point of view is.

Terminological note

Representation is one of the many philosophical terms that have very long histories and a lot of different meanings. Almost every philosopher speaking about representation has interpreted it in a particular way. Philosophers of knowledge, political philosophers, philosophers of law, philosophers of art and so on have all used it, and sometimes without great philological caution as to its many meanings. Moreover, in particular in the fields I am interested in, that is, epistemology and the philosophy of science, the history of *representation* has often intersected with the history of two other related and problematic terms: *image* and *picture*. As if that were not enough, the translation from the old Greek to the Latin, and from the Latin to the other philosophical languages has also created enormous problems. In addition there are the difficulties of translation from one contemporary philosophical language to another. It would be enormously difficult to reconstruct the history of such a term. However, a warning on the inextricable mass of meanings concerning *representation* – highlighting a few points – can be worthwhile, even if limited to English and German. Note, however, that this does not mean that I want to propose a philosophy of representation based on the etymon of the term. This philosophical path, as is well known, is actually a blind alley.

In English we have *representation, image, picture,* and often these terms are considered as synonymous. In German we have *Bild* and cognates (*Bildung, Abbildung, Einbildung* and so on), *Vorstellung, Darstellung,* which have also often been considered as synonymous. All of them are linked with Plato's doctrines of *phantasía* and *eikasía,* that is, with what is probably the first explicit theorization concerning representation. Moreover both the English and German terms have to do with the medieval reconsiderations of this theorization, and therefore with terms like *imago* and *repraesentatio,* which have been philosophically explored by Augustine, Boetius, and then magisterially by Thomas Aquinas. The problem was, along with the birth of the modern philosophical languages, how the Greek and the Latin terms should be translated. Often the old root has been maintained, as in the neo-Latin languages and in some terms in the Anglo-Saxon languages. But sometimes the appearance in the philosophical scenario of a term has been more complicated. Let us think of the German '*Bild*'. It was suggested as a translation of the Greek *eidolon, eikon,* and the Latin *imago, species,* and probably derives from the old German *bilidi,* that meant something like 'representation endowed with magic power'. The same happened for the German *Vorstellung,* which was already available to the first translators into German of the Greek and Latin philosophical texts, and which was used to translate the Greek *phantasía* and the Latin *repraesentatio*.

But things are further complicated by the mutual translations between contemporary languages. Let us look at a couple of examples from two German philosophers – Frege and Wittgenstein – whose works provide the basis for many contemporary Anglo-Saxon ways of thinking.

First let us consider Frege and his 'Thoughts' (Frege, 1918–19; German text in Frege, 1969) and compare some English passages with the original German ones (my italics).

We find truth predicated of *pictures, ideas,* sentences, and thought. (p. 2)	Man findet die Wahrheit ausgesagt von *Bildens, Vorstellungen,* Sätze und Gedanke.
A *picture* is meant to *represent* something. (Even an *idea* is not called true in itself, but only with respect to an intention that the idea should correspond to something.) (p. 3)	Das *Bild* soll etwas *darstellen.* Auch die *Vorstellung* wird nicht an sich wahr genannt, sondern nur im Hinblick auf eine Absicht, dass sie mit etwas übereinstimmen solle.
Now do thoughts belong to this inner world? Are they *ideas*? (p. 14)	Gehören nun die Gedanke dieser Innerwelt an? Sind sie *Vorstellungen*?

Here, *Bild* has been translated by *picture,* and *Vorstellung* by *idea.* We know, by reading Frege's paper, that he intended to give a subjective connotation to *Vorstellung,* contrary, for example, to Kant's use of the same term, which, by the way, is usually translated by *representation.* Frege writes that a *Bild* 'represents something', and he uses the verb *darstellen.* This is in the tradition, accepted by some phenomenologists, of using *Vorstellung* to indicate a subjective representation, and *Darstellung* to indicate an objective representation.

Let us now consider Wittgenstein. We know that in his 1922 *Tractatus logico-philosophicus* there is the shadow of Hertz's 1894 masterpiece, *The Principles of Mechanics* (compare, for example, Barker, 1980; Tougas, 1996). By reading the *The Principles* we realize that Hertz uses the term *Bild* in at least three different senses: (1) as a generic representation of the phenomena; (2) as a theory; and (3) as a particular interpretation of a theory. Taking this into account, let us compare the English and the German versions of some passages of the *Tractatus.*

We make ourselves *pictures* of the facts. (2.1)	Wir machen uns *Bilder* der Tatsachen.
The *picture* is a model of reality. (2.2)	Das *Bild* ist ein Modell der Wirklichkeit.

The *picture represents* what it *represents*, independently of its truth or falsehood, through the form of *representation*. (2.22)	Das *Bild stellt* dar, was es *darstellt*, unabhängig von seiner Wahr- oder Falschheit, durch die Form der *Abbildung*.

Note, by the way, with reference to Wittgenstein's statement 2.1, that there is a very similar passage by Hertz in the Introduction to his 1894 book: 'Wir machen uns innere *Scheinbilder* oder Modelle der Gegenstände', or 'We form for ourselves images or symbols of external objects' (Hertz, 1894, p. 1).

Both Hertz and Wittgenstein speak about *Bild* and about *darstellen*. But the English translation differs: Hertz's *Bild* is translated by *image*, while Wittgenstein's *Bild* is translated by *picture*. In both the translations the term *represent* occurs, although this tends to underestimate the slight difference between the German *vorstellen* and *darstellen*.

While the linguistic complexity of *representation* may prove inextricable, in the field of the theory of knowledge, all the terms above, and the different doctrines of representation, share two aspects which follow from their origin to the present day. First, representation can be thought of either as an act or as a product of that act: the former relates to psychology and cognitive sciences; the latter to epistemology. In what follows I focus on the latter. Second, any representation is something that represents something, and between the representation and the represented there is some kind of similarity depending on the particular doctrine of representation we are dealing with. In this book, representation is both *id quod conoscitur*, that is, what we know, and *id quo conoscitur*, that is, that by means of which we know by cognitively constituting the world in which we live and act. This means that will I defend neither an isomorphic theory of representation, according to which the representation and the represented share the same elements and the same relations among the elements, nor a mirror theory of representation, according to which the latter mirrors both what there is in the world and what occurs in it.

Philological note

With reference to Kant's writings, I use the English translations indicated in the bibliography. However I have compared them with I. Kant, *Werkausgabe* (1956–64). In particular, I quote:

- *Kritik der reinen Vernunft*, as (KdrV, p. xx, Byy);
- *Prolegomena zu einer jeden künftigen Metaphysik, die als Wissenschaft wird auftreten können*, as (P, p. xx, yy);
- *Metaphysische Anfangsgründe der Naturwissenschaft*, as (M, p. xx, yy);
- *Kritik der Urteilskraft*, as (KdU, p. xx, yy);
- *Logik*, as (L, pp. xx).

Where xx indicates the pages of the English translation used, and yy the pages, if quoted, of the original text to which the translation refers. Note that, except on very few occasions, I do not use the first edition of the *Kritik der reinen Vernunft* (this is why in the notes I quote only the B version). I believe that Kant, in proposing his second edition, intended to offer the reader, for reasons I do not deal with here, a version more correspondent to his real thought and less open to unacceptable interpretations. Finally, I do not consider the *Opus postumum*. In all honesty, it is too difficult to find a real path to disentangle its content or to make a good argument for it.

1
Concepts and Objects

'What is a concept?': our hard and *vexata quaestio*! With good will and patience we could consider all the answers thrown up by a look at the history of philosophy, one by one, in order to find the best one, that one that satisfies our own personal philosophical feeling. Although this does not really seem such a good idea: we would probably be dead long before we arrived at the end of the list of possible theories. To avoid this inane and insane effort, I will focus on only four of them, and then I will propose my own. The leading thread that we will follow will be based on Cassirer's proposal to fuse Kant's doctrine of concept with Frege's. In so doing, especially in *Substance and Function*, he offers a very stringent critique of Aristotle's classical theory of concept. By means of his extremely lucid analysis it will be clear that there is a wide conceptual gap between the classical theory, which considers *concepts as classes* (class-concepts), and the modern and contemporary theories, which consider *concepts as functions* (function-concepts).

First, I discuss the main features of Aristotle's, Kant's, Frege's and Cassirer's proposals. From a historical-critical viewpoint I aim to (1) show their weakness, (2) underline possible terminological ambiguities (in particular that one related to the notions of 'function' à la Kant and 'function' à la Frege), and (3) explicitly state the tradition into which my proposal fits (that is, a Kantian tradition). Next, I present my own thesis. The main point of my proposal concerns the idea that the concept, intended as a knot of more basic concepts, the *characteristics*,[1] can be thought of both as representation and as rule. Alongside this point, a theory of objects, regarded as a knot of properties, will also be proposed.

In the third part, the notion of 'reality' will be discussed. Reconsidering the question of realism today seems useful, since beyond those who plead for the death of God, of art, of scientific method, of philosophy as a whole, there are also those who plead for the death of reality. In particular, there are those who consider reality completely dissolved into scientific representation. Others, in the name of a rough and a-philosophical instrumentalism, sometimes disguised as a taught pragmatism, think the problem of realism totally negligible, or solvable by resorting to the engineers.

With reference to the former – the apocalyptics – a short paper by Poincaré (1906), emblematically titled 'La fin de la matière', should be recalled. Here, the French physicist-philosopher discusses, ironically enough, W. Kaufmann's comments on his two 1902 experiments on the ratio between the charge and the mass of the electrons. According to Kaufmann, the direct consequence was the dissolution of inertia into the electromagnetic field, and thus the end of material objects. Poincaré addressed to Kaufmann and the supporters of the dissolution of reality into the last coeval scientific representation the same invitation to prudence and encouragement to wait for new theoretical and experimental results that, nowadays, should be directed towards the new theorists of the 'end of matter' in the name of, for example, field theory or quantum gravity.

With reference to the latter – the instrumentalists – it could be noted that even if instrumentalism were a good approach to settling certain practical problems, it could not be said that it is always a profound way out for any philosophical problem.

1.1 The descent of substance and the ascent of function

As recalled, Cassirer, on the one hand proposes a new theory of concept and on the other radically criticizes the classical theory according to which a concept is thought of as a class. This theory was formulated within Greek philosophy and had its climax during the Middle Ages, when Aristotelian logic was rewritten more systematically. In fact, it is Aristotle himself who is regarded by Cassirer as the paradigmatic supporter of such a theory.

The theory of concept as a class is based on the idea that, given a set of objects, human thought, by comparing them and reflecting on the comparison, isolates common features, which in turn become the concept. For example, through comparison of and reflection on a set of objects, such as beeches, evergreen oaks, firs, larches and so on, human thought discovers what is common to them and abstracts the concept of 'tree'. Similarly, by comparing mathematical objects, such as rhombi, squares, trapezes and rectangles, and then by reflecting on what is common, it abstracts the concept of 'quadrilateral'. This process of comparison, reflection and abstraction can be further refined. For example, by comparing trees, shrubs, flowers and so on, and by reflecting on what is common, the concept of 'vegetable' is abstracted. By comparing quadrilaterals, triangles, pentagons, hexagons, and so on, and by reflecting on what is common, the concept of 'plane figure' is abstracted.

By means, therefore, of a process of continuous comparison, reflection and abstraction, thought forms more and more general concepts. In this way concepts subsume more and more objects (that is, their extension is larger and larger); at the same time, concepts need fewer and fewer characteristics (that is, their comprehension is smaller and smaller). Therefore this tradition individuates the concept extensionally, as the class containing all the objects having a particular common property; that is, we have a class-concept.

Note that there is a difference between *abstracting what is common* (*abstrahere aliquid*) from *abstracting from what is not common* (*abstrahere ab aliquo*). Yet Cassirer seems to think that the classical theory of concept considers both aspects, because thought both abstracts what is common by considering it in the concept, and abstracts from what is not common by not inserting it into the concept.

However, Cassirer claims that a good theory of concept should satisfy two necessary requirements: (1) it must allow us to define precisely what we are talking about; (2) it must give an account of all the scientific concepts. Classical theory does not fulfil both these requirements.

Were the process of formation of concepts really an abstraction from what is not common to what is common, it would be a process by denial, and it would lose what is denied. Furthermore, how can we be sure what is not included is inessential? With regard to this point, Cassirer recalls H. Lotze's ironic example. Both cherries and beef are red, edible and juicy. If we abstract what is common and deny what is not common between cherries and beef, we should obtain the concept of 'red, edible and juicy'. Unfortunately, this concept is absolutely undetermined, and it does not allow us to assert anything significant. Furthermore, we cannot go back from it to cherries and beef, since what characterizes those objects *qua* those objects has been lost in the process of the concept formation.

On the contrary, the concept must tell us precisely what falls under it. Moreover, if it is a high-level concept, it must let us identify precisely also the derivative concepts, that is, the species-concepts, of which it is the genus-concept. In conclusion, the logical criticism of the classical theory of the concept formation targets the fact that: 'the concept would lose all value if it meant merely the neglect of the particular cases from which it starts, and the annihilation of their peculiarity' (Cassirer, 1910: p. 6). However, within the classical tradition – as Cassirer concedes – Aristotle's position should be distinguished from the others because here the class-concept is not just a logical wrapping containing objects with common properties. Rather, it has also an ontological value:

> [it is] the real Form which guarantees the causal and teleological connection of particular things. The real and ultimate similarities of things are also the creative forces from which they spring and according to which they are formed [...] The determination of the concept according to its next higher genus and its specific difference reproduces the process by which the real substance successively unfolds itself in its special forms of being. Thus it is this basic conception of substance to which the purely logical theories of Aristotle constantly have reference. The complete system of scientific definitions would also be a complete expression of the substantial forces which control reality.
>
> (Ibid.: pp. 7–8)

Consequently, the *Aristotelian class-concept is also a substance-concept*. It would, therefore, avoid the criticisms considered above only if the Aristotelian theory of substance remains valid. It seems that the Aristotelian theory of substance-concept would work in the realm of natural sciences, since it would seem to offer an ontologically grounded classification of species and genera. Nevertheless, its weaknesses are exposed once we cease to deal not with the 'form of the olive-tree, of the horse, of the lion' but with mathematical objects, especially geometric ones. How can concepts such as 'point', 'line' and so on, fit into the Aristotelian framework? What are they abstracted from? What are the common characteristics of more primitive objects which have been denied? What are the common properties shared by them? Neither the more ontologically grounded substance-concept conception, nor the less ontologically grounded class-concept conception appears to offer a satisfying solution.

Note that Cassirer's primary critical target is the classical doctrine of class-concept, and only secondarily the Aristotelian doctrine of the substance-concept, which is an ontological instantiation of the former. Actually, the same criticisms can be addressed to another possible instantiation of the classical theory: the psychologistic version supported, for example, by Berkeley. To put it another way, Cassirer does not want to address his criticisms specifically to Aristotelian conceptual realism or to Berkeleian nominalism, but rather to the logical structure they share. The target of his criticism is not the possible epistemological and ontological interpretations of the theory of class-concept, but the logical structure on which it is based. That is, he wants to challenge the idea that concepts are abstractions of common properties, and its consequences, that is, that determining a Porphyry-tree structure for each succession of abstracting concepts is possible. However, Aristotle's original theory was much more carefully articulated than Cassirer's reconstruction in *Substance and Function*, and it is worth recalling it in more detail.

1.1.1 Some notes on Aristotle's theory

Facing Aristotle means plunging into a very deep and dangerous sea in which to swim and not to drown is extremely hard for a non-philologist. It is a sea made up of interpretations, counter-interpretations, counter-counter-interpretations and so on. And each of these is based on an erudite reading of the entire *corpus aristotelicum*. However, I will skip this kind of analysis and limit myself to recalling some points of Aristotle's theory of concept, rather of his theory of categories.

I wish to begin my survey from Kant's criticism:

> It was an enterprise worthy of an acute thinker like Aristotle to make search for these fundamental concepts [the categories]. But as he did so on no principle, he merely picked them up as they come in his way, and at first procured ten of them, which he called *categories* (predicaments).[2] Afterwards he believed that he had discovered five others, which he added under the name of post-predicaments. But this table still remained defective. Besides, there are to be found in it some modes of pure sensibility (*quando, ubi, situs*, also *prius, simul*), and an empirical concept (*motus*), none of which have any place in a table of the concepts that trace their origin to the understanding. Aristotle's list also enumerates among the original concepts some derivative concepts (*actio, passio*), and of the original concepts some are entirely lacking.
>
> (KdrV, p. 114, B107)

Kant's severe judgement is certainly not benevolent. In brief, Kant accuses Aristotle (1) of being rhapsodic, (2) of deriving categories from experience, (3) of also considering concepts that are not categories,

(4) of listing prime concepts as well as derivative concepts and (5) of not producing a complete table. These five imputations have different natures, and to understand them is to consider the possibility of whether Aristotle is justly accused.

The first and the fifth concern the Aristotelian proposal as such, which is considered to lack a leading thread and not to be exhaustive. Actually this is true, as we will see. The second accusation deals with the Aristotelian set-up and not with the list of categories. That Aristotle derives his table from experience and not from an a priori principle – as Kant will do – cannot be considered as a fault. It is a consequence of the Aristotelian conception as a whole, and if criticism is due it should be the latter that is criticized and not one of its consequences (that is, a particular aspect of the theory of category). Finally, the third and the fourth accusations concern the different position of some categories in Aristotle's theory in comparison with the position they hold in Kant's. Here we have two different philosophical conceptions. And, of course, it appears rather obvious that some elements can have different roles within the two different theories. It seems that Kant is accusing Aristotle of not thinking as he – Kant – thinks.

On categories

Does the history of the notion of 'category' begin with Aristotle, or should its origin be traced back to the dawn of philosophical reflection? It is impossible to settle this question once and for all and there are different possible answers depending on the particular historical interpretation we share. If the logical aspect of the theory of category is privileged, then its history can be made to start, as Trendelenburg (1846, pp. 196–202) proposes, with Socrates, who was, so to speak, the 'discoverer' of the universality of concepts, that is, of its applicability to many subjects. If, instead, the ontological aspect is privileged, then its history can be dated, as Ragnisco (1871, pp. 61–72) suggests, back to those times in which categories were still hidden in myth.

Two different interpretations, therefore, allow us not only two different readings of the same philosophical theory, but two different estimations of their birth date as well. Trendelenburg and Ragnisco were great and skilful historians of philosophy and surely we cannot simply claim that one is wrong and other is right. Actually they searched for two different things. Trendelenburg searched for the time when a logical use of categories – even if unconscious – occurred for the first time in the history of philosophy. Ragnisco searched for the time when a conceptualization of becoming, and therefore, a conscious use of category, occurred for the first time.[3]

Already we can see that identifying the real definition of 'category' in Aristotle's works is highly complex and depends on the particular hermeneutics through which we read them. However, there is a point on which all the historians agree: Aristotle was the first to attempt a detailed theory of concepts, rather, of categories. Indeed, we owe the introduction to philosophical debate of the term 'category' to Aristotle. Before him, the term was used in juridical language to mean 'accuse', that is, as something opposed to 'apologia'. Porphyry, in his commentary on the Aristotelian theory of categories, observed that Aristotle used the term for the first time to indicate those locutions that impute something to be something. For example, claiming that 'This is a stone' means imputing that thing to be a stone.[4] Note that the Greek term was then translated by Boetius (liber I, p. 75) by 'predicament', even if it is not so certain that all the Aristotelian categories have the same predicative role, or the same epistemological status.

Let us stay with the interpretative problems concerning the definition and let us recall that there are at least two different factions supporting two different definitions of the term 'category'. The first one is based on the logico-grammatical interpretation of the Aristotelian theory of categories: categories are 'things said without any combination' (*Cat.* 4, 1b 25).[5] The second is founded on the ontological interpretation: categories are connected to the senses of 'being': 'those things are said in their own right to be that are indicated by the figures of predication: for the senses of "being" are just as many as these figures' (*Metaph.* V, 7, 1017a 20–5).[6] Between these there are, of course, other interpretations. In particular, there are scholars who claim that the logico-grammatical and the ontological definitions are the two necessary faces of the same coin (compare Ross, 1924; Vollrath, 1969).

I have spoken about the theory of the highest concepts (that is, categories), but in Aristotle there is also a specific doctrine of concepts. In *De Interpretatione*, next to an analysis of apofantic discourse, there is an analysis of the elements composing that discourse, that is, subject and predicate. Those interpreters of the Aristotelian theory of categories supporting the logico-grammatical approach emphasize precisely this analysis. Those supporting the ontological approach claim, on the contrary, that it is not important, even disputing Aristotle's authorship of some of the writings – for example, *Categoriae* (see Ragnisco, 1871). Those suggesting that logic-grammar and ontology are two sides of the same coin argue that what Aristotle writes in *Categoriae* and in

De Interpretatione is also to be found in *Metaphysics*, even if the point of view is different.

According to this third interpretation, which Cassirer seems to share, the analysis of propositions developed at the beginning of *De Interpretatione* cannot be taken to mean that language is somehow fundamental. It – language – is only a tool useful to identify the real fundament, that is, ontological aspect. As far as this point is concerned, Aristotle himself suggests this hierarchy elsewhere by claiming that truth and falsehood have firstly to do with what is 'in the soul', and only subordinately with what is 'in spoken sounds' (*De Int.*, 1, 16a 9–11).[7] Thus, even if we start from 'speech' – that is, language – this is not the fundament. Language is a medium allowing us to reach the real ontological fundament. That is, according to Aristotle (even if relating to a different question), it is natural,

> to start from the things which are more knowable and clear to us and proceed towards those which are clearer and more knowable by nature; for the same things are not knowable relatively to us and knowable without qualification. So we must follow this method and advance from what is more obscure by nature, but clearer to us, towards what is more clear and more knowable by nature.
>
> (*Phys.* I, 1, 184a 16–22)

This point should be kept in mind when we come to deal with Frege, since we will meet the same problem about priority, and more or less in the same terms.

Categories and predication

In section 4 of *Categoriae*, Aristotle lists the ten highest genera: substance, quantity, quality, relation, place, time, being-in-a-position, having-affected, doing-affected, being-affected.

> To give a rough idea, examples of substance are man, horse; of quality: four-foot; of qualification: white, grammatical; of a relative: double, half, larger; of where: in the Lyceum, in the market-place; of when: yesterday, last-year; of being-in-a-position: is-lying, is-sitting; of having: has-shoes-on, has-armour-on; of being-affected: being-cut, being-burned.
>
> (*Cat.* 4, 1b 25–2a 5)

But has Aristotle considered all the possibilities? With reference to the number ten, it is supposed that it had a Pythagorean origin, and therefore that it should be thought of more as a numerological vestige than as something having ontological value. However, Aristotle does not restrict himself to this number: sometimes he indicates ten categories, sometimes eight, sometimes six (see Ragnisco, 1871, pp. 295–6). Moreover, according to some scholars, sometimes Aristotle also indicates as categories concepts which do not belong to the original list, such as 'masculine', 'feminine', 'walking', 'length', 'largeness', 'number', 'colour', 'form', 'matter', 'possible', 'necessary', 'flying', 'swimming' and so on (Ibid.). Certainly the number ten does not seem so relevant from the theoretical point of view. In the *Corpus aristotelicum*, besides *Categoriae*, where it is not explicitly mentioned but ten categories are listed, it is explicitly stated only in the *Topica*:

> Next, then, we must distinguish between the categories of predication [...] These are *ten* in number: What a thing is, Quantity, Quality, Relation, Place, Time, Position, State, Activity, Passivity.
>
> (*Top*. I, 9, 103b 20–5, emphasis added)

Note that the non-exhaustiveness of the table of categories can be justified also by resorting to the particular architectonic of the Aristotelian system. While Kant derives them by a priori considerations, Aristotle realizes the list by resorting both to an a priori moment, connected with the form determining the otherwise undetermined sensible matter, and to an a posteriori moment, since the modes of being of the empirical data must be taken into account. It is exactly this a posteriori moment which indicates that the table cannot be exhaustive: the empirical infinite multiplicity cannot be subsumed in its entirely into the few categories mentioned.

However, let us consider the central point of the list, that is, the connection between the first category, that of substance, and the other categories. It is a division – probably derived from Plato – which fixes an ontological priority:

> Now there are several senses in which a thing is said to be primary; but substance is primary in every sense – in formula, in order of knowledge, in time. For of the other categories none can exist independently, but only substance. And in formula also is primary; for in the formula of each term the formula of its substance must be present. And we think we know each thing most fully, when we know

> what it is, for example what man is or what fire is, rather than when we know its quality, its quantity, or where it is; since we know each of these things also, only when we know *what* the quantity or the quality *is*. And indeed the question which, both now and of old, has always been raised, and always been the subject of doubt, viz. what being is, is just the question, what is substance? For it is this that some assert to be one, other more than one, and that some assert to be limited in number, other unlimited. And so we also must consider chiefly and primarily and almost exclusively what that *is* which *is* in this sense.
>
> (*Metaph.* VII, 1, 1028a 30–1028b 5)

Therefore, on the one hand, there is the substance and, on the other hand, there are the fundamental rubrics, the highest genera, containing the determinations predicated, as accidents, of what is subsumed in the first one: 'the things which do not signify a substance but are said of some other underlying subject which is neither just what is that thing nor just what is a particular sort of it, are accidental, for example white of the man' (*An. post.* I, 22, 83a 25–30). This second group indicates the possible modes according to which the substance can exist: as endowed with a certain quality and a certain quantity, as being in a certain time and in a certain position, as acting, and so on. Nevertheless, substance is the category without which the other categories would not have any value: we cannot speak of quantity, quality, time, position, action and so on, without something which can have such determination, that is, without substance.

Note that the *accidental* categories, being schemata of empirical accidents, are not accidents in the usual sense. For if being 'white', or 'good' can pertain to man accidentally, since a particular man can be white, or good; it is not at all accidental that a man has the possibility of having a quality such as 'whiteness', or 'goodness'.

To enable a better comprehension of the logical aspect of categories, it is worth looking at the theory of predication. This is based on the distinction between *saying of* and *being in*, that is, between *predication* and *inherence*. Predication is saying something of a subject, and therefore it is proper to both essential predicates and accidental predicates. Inherence is *being in a subject*, not in the sense of an essential part of a subject but as something which cannot exist as such. Both predication and inherence concern the first category, since the substance is the subject of any predication (logical aspect), and the substrate of any inherence (ontological aspect).

On the basis of this distinction, Aristotle, in *Categoriae* (2), differentiates four classes:

(1) *What is said of a subject, but is not in a subject.* For example, 'man' is said of a subject (a certain man), but is not in any subject. This is the characterization of the secondary substances,[8] that is, of the essential predicates of the subject. Here we are dealing with the concepts belonging to the same category of the subject.
(2) *What is in a subject, but it is not said of a subject.* For example, a given white is in a subject (in what is white), but is not said of any subject. This is the characterization of the individual accidents, that is, of the individual predicates that do not belong to the same category of the subject.
(3) *What is said of a subject and is in a subject.* For example, science is in a subject, that is, in the soul, and it is said of a subject, that is, science-of-grammar. This is the characterization of the universal properties.
(4) *What is not in a subject, and neither is said of a subject.* For example, 'a determined man' and 'a determined horse' are not said of a subject, and they are not in a subject. This is the characterization of the real substance.

From this, it should be clear that the predication can occur:

(1) among concepts belonging to the same category (for example, 'man is an animal', where 'man' and 'animal' belong to the same category), and therefore we have the essential predication;
(2) among concepts belonging to different categories (for example, 'Socrates is white', where 'Socrates' and 'white' belong to two different categories), and therefore we have the accidental predication.

Let us note that while the primary and the secondary substances are characterized by not being in a subject, the accidents are characterized by being in it.

	Saying of	*Not saying of*
Being in	Universal properties	Individual properties
Not being in	Secondary substances	Real substance

The Aristotelian theory of predication is completed with a discussion of the modes according to which a concept can be predicated of a subject.[9] There are five such modes:

(1) *genus*, when it indicates the subject generically and indeterminately; for example, 'Socrates is an animal';
(2) *species*, when it indicates the subject specifically and determinately; for example, 'Socrates is a man';
(3) *difference*, when it indicates the difference distinguishing that species from other species of the same genus; for example, 'Socrates is rational';
(4) *proper*, when it indicates something necessary to the essence of the subject, even if it is not essential in itself; for example 'Socrates laughs';
(5) *accident*, when it indicates a trait that could also not be; for example, 'Socrates is bald'.

While genus, species, and difference are essential modes of predication, proper and accident are not. However, attention should be paid to the fact that 'accident' in a logical sense must be differentiated from 'accident' in an ontological sense, that is, *accident as predicable* from *accident as predicament*. For the accident as category is opposite to substance (ontological sense); but the accident is also a mode according to which a concept can be predicated of a subject (logical sense). Therefore, only 'accidentally' the two aspects coincide. For example, 'rational' is an accidental concept from the ontological point of view, but from the logical point of view it is predicated as difference; while 'good' is accidental both as predicament and as predicable.

The principle of predication, according to which what can be predicated of a predicate can be predicated also of the subject of that predicate (*nota notae est nota rei ipsius*), is important, since it tells us that a *nota* (that is, a characteristic) of a concept X is also a *nota* (that is, a characteristic) of the concept, or of the object, Y, of which X is a *nota* (that is, a characteristic). Thanks to the theory of categories, Aristotle also offers a definition of this:

> Whenever one thing is predicated of another as of subject, all things said of what is predicated will be said of the subject also. For example, man is predicated of the individual man, and animal of man; so animal will be predicated of the individual man also – for the individual man is both a man and an animal.
>
> (*Cat.* 3, 1b 10–15)

Let us note here that there are some ambiguities, due to the fact that the two notions of 'concept' and 'object' are not yet correctly distinguished. For Aristotle does not say anything about the fact that the subject can be both a concept and an object. Moreover, this ambiguity leads towards other ambiguities such as that between 'characteristic' and 'property' (and their correlates, 'attribute', 'quality', 'trait', 'moment'). These ambiguities will be partly eliminated only when what belongs to an object (properties) and what belongs to a concept (characteristics) will be logically, epistemologically and ontologically distinguished by Kant and Frege. However, the ambiguity between 'property' and 'characteristic' is not simply due to a lack of distinction between 'concept' and 'object', but also to the desire to make the two notions ontologically equivalent. Of course this is also connected with the problem of universals, and to their possible realist interpretation. In this case, there is no great difference between the reality of a concept (that is, a universal) and the reality of an object, and it becomes possible to use either 'property' or 'characteristic' in both cases (the property of a concept, the characteristic of an object).

Coming back to the definition according to Aristotle, defining a term means giving the essence of what it refers to, that is, it means giving the 'what it is', or determining the species to which what is referred belongs. For to define a term is sufficient to determine the proximal genus and the specific difference which distinguishes that species from the other species of the same genus. While the genus determines the essence in an undetermined way, the difference focalizes the essence by indicating the species.

Therefore, defining a term means inserting it into the right position of the Porphyry-tree (compare *Isagoge*, pp. 110–12), that is, into that point of the hierarchical tree-structure that is immediately beneath the proximal genus and that belongs to that determined species. For example, 'man' is defined as 'rational animal', where 'animal' refers to the proximal genus and 'rational' refers to the specific difference allowing us to determine the right species by characterizing it.

highest genus
(substance)
...
...
subordinate
genera and species
(vertebrate, animal, rational)

...

...

lowest species

(man)

individuals

(Plato, Socrates, Callias)

Under the lowest species, expressing the last universal determinations belonging to more than one individual, there are the individuals that are differentiated from each other by accidental properties. It follows that the individuals cannot be defined, since they can neither be predicated nor expressed through concepts, because these are universal. Individual can be only described.[10] But the highest concepts cannot be defined either, since they are the highest epistemological level and there are no concepts above them representing their proximal genus.

> Of all the things which exist some are such that they cannot be predicated of anything else truly and universally, for example Cleon and Callias, that is, the individual and sensible, but other things can be predicated of them (for each of these is both man and animal); and some things are themselves predicated of others, but nothing prior is predicated of them; and some are predicated of others, and yet others of them, for example man of Callias and animal of man. It is clear then that some things are naturally not said of anything; for as a rule each sensible thing is such that it cannot be predicated of anything, save incidentally – for we sometimes say that white object is Socrates, or that that which approaches is Callias. We shall explain in another place [*An. post.* I 19–22] that there is an upward limit also to the process of predicating; for the present we must assume this.
>
> (*An. pr.* I, 27, 43a 25–40)

The Porphyry-tree is, therefore, the hierarchical structure which, starting from the individual, rises through increasing universality, up to the highest genus. And this logical structure is exactly what was criticized by Cassirer.

1.1.2 Moving to Kant's theory

While discussing Cassirer's criticisms, I have emphasized that he was under Kant's intellectual influence. Now is the time to recall some aspects of this influence.

Let us begin with the problem of what a category is. As we have seen, according to Aristotle, categories are the highest genera. This is not true in the Kantian conception. So what are categories here? Several times, Kant claims that they are 'logical functions', which are needed to give 'unity to the mere synthesis of various representations *in an intuition*' (KdrV, p. 112, B104). Thus, they are functions; but what does that mean?

> By 'function' I mean the unity of the act of bringing various representations under one common representation.
>
> (Ibid., p. 105, B93)

In other words, a function is the synthesis produced by the activity of the understanding. Therefore, the understanding is not passive, but spontaneous, and it connects representations by unifying them into one. Consequently, concepts and, a fortiori, categories, are functions of synthesis by means of which the knowing subject judges, that is, produces judgements that are representations of representations. Thanks to the fact that knowing, and therefore thinking, is judging, to obtain the table of categories from the table of judgements is possible. This is the Kantian *Leitfaden*:

> Thought is knowledge by means of concepts. But concepts, as predicates of possible judgments, relate to some representation of a not *yet* determined object. Thus the concept of body means something, for example metal, which can be known by means of that concept. It is therefore a concept solely in virtue of its comprehending other representations, by means of which it can relate to objects. It is therefore, the predicate of a possible judgment, for instance, 'every metal is a body'. The functions of the understanding can, therefore, be discovered if we can give an exhaustive proposition of the functions of unity in judgments.
>
> (Ibid., p. 106, B94)

Let us note, in this passage, a very important aspect which will be met in Frege too: the priority of judgements over concepts, that is, the fact that concepts are 'predicates of possible judgments'. Another important point to note is that concepts, especially the pure concepts (that is, the categories), are no longer taken extensionally (as class-concepts), but intensionally (as function-concepts). In this way, a fundamental turning point in the theory of concepts is realized. However, let us come back to the definition of 'category' which Kant seems to give in the extract above. Yet, Kant resolutely rejects such a possibility:

> we cannot define any of them [the categories] in real fashion, that is, make the possibility of their object understandable, without at once descending to the conditions of sensibility, and so to the form of appearances – to which, as their sole objects, they must consequently be limited. For, if this condition be removed, all meaning [significance], that is, relation to the object, falls away; and we cannot through any example make comprehensible to ourselves what sort of thing is to be meant by such a concept.
>
> (Ibid., p. 360, B300)

Observe that he states the *impossibility of the definition of individual categories*. To make this point clearer, we need to bear in mind what followed this passage in the 1781 edition of the *Critique* only.[11] Here Kant explains that he does not give any definition of the individual categories, since not only is it not necessary for the sake of his argument, but also for 'precaution', that 'lies still deeper. We realize that we are unable to define them even if we wished' (Ibid., p. 261, A241). To this problem, and still on the same page of the 1781 edition, the following footnote is dedicated:

> I here mean real definition – which does not merely substitute for the name of a thing other more intelligible words, but contains a clear property by which the defined object can always be known with certainty, and which makes the explained concept serviceable in application. Real explanation would be that which makes clear not only the concept but also its objective reality. Mathematical explanations which present the object in intuition, in conformity with the concept, are of this latter kind.
>
> (Ibid., p. 261, A241, fn. A)

Even if we wanted a *real definition*, we would not obtain it because: (1) we are analysing the categories in themselves; and (2) what gives reality to the objects, by means of sensation, is the empirical intuition (compare Ibid., p. 184, B182–3). Since to give a *real definition* to something that does not contain any element of reality is impossible, we cannot define individual categories, for they must be considered separately from the empirical intuition (Ibid., pp. 262–3, B302).[12]

Categories in a pure sense are 'concepts of an object in general [*Gegenstande überhaupt*]' (Ibid., p. 128, B128), that is, they have nothing to do with any particular object given by the synthesis of the actual sensible intuitions. In this sense, they are forms of experience in 'general' [*Erfahrung überhaupt*], which cannot give us any empirical knowledge (Ibid., pp. 252–3, B288).[13] This means that, on the one hand, categories are not sufficient to know objects, which always require an empirical intuition, but, on the other hand, they are the necessary conditions for such knowledge, since they are the transcendental forms which make that act cognitively possible (Ibid., p. 171, B161). Let us observe that even if these remarks cannot be used to obtain a real definition of each individual category, they can be used to obtain a different kind of definition, in particular what I call a *transcendental definition: categories are the formal conditions of any possible experience*. There is something more. A category, both in its general and its individual sense, can be considered in relation to its aim, that is, as the *logical function unifying the synthesis of representations*. This is not the substitution of a name by different words, as Kant feared, but it can be considered as a genuine definition, in particular what could be called an *operational definition*: we know what categories are, since we know what their use, their function, is.[14]

The transcendental definition concerns only categories, whereas the operational definition can be given both to categories (pure concepts) and to empirical concepts. All of these are logical functions unifying the synthesis of the manifold, that is, they are rules of synthesis.

While pure concepts primarily concern understanding, empirical concepts primarily concern experience. Precisely because of the connection with experience, the universality of empirical concepts is wholly contingent and can never become necessary. Only if we bear this in mind, do we not find a complete inconsistency between what Kant writes in the *Critique* and what he writes in the *Logic*.[15] In the *Logic* there is a theory of empirical concepts, but it is strongly influenced by the classical approach. Here he writes that an empirical concept is a universal (*per notas communes*) or selective (*discursiva*) representation, and it is opposed to intuition, which is an individual (*singularis*) representation (L, p. 96).

In the Kantian logic textbook, then, we can find real classical themes. Kant notes that general logic cannot explain the genesis of empirical concepts, since it does not deal with objects: this can eventually be pursued by psychology. Furthermore, logic cannot even explain the genesis of concepts in relation to matter, that is, their being empirical, or arbitrary, or intellectual: this can be the concern of metaphysics.

Yet, from the point of view of their form, one can investigate 'how given representations become concepts in thinking' (Ibid., p. 99), that is, how representations can be subsumed (not unified) under a common representation. This can be realized by means of three logical acts:

(1) *comparison* ('the likening of representations to one another in relation to the unity of consciousness');
(2) *reflection* ('how [different representations] can be comprehended in one consciousness');
(3) *abstraction* ('the segregation of everything else by which given representations differ').

> For example I see a fir, a willow and a linden. In firstly comparing these objects, I notice that they are different from one another in respect of trunk, branches, leaves, and the like; further, however, I reflect only on what they have in common, the trunk, the branches, the leaves themselves, and abstract from their size, shape, and so forth; thus, I gain a concept of a tree.
>
> (Ibid., p. 100)

There are two remarkable points here. First, what Kant proposes and how he exemplifies it could have been given by a Scholastic to support his thesis about class-concepts formation. Nevertheless, this has to be considered carefully: a scholastic could never have defined the three acts in the same way. Kant's definition contains a reference to the 'unity of consciousness', which allows us to interpret them in a (almost-) transcendental way.[16] Second, Kant means 'abstraction' in the sense of separation. He does not speak about abstracting something (*abstrahere aliquid*), but about abstracting from something (*abstrahere ab aliquo*) (Ibid., p. 101).

Nevertheless, *Logic* mirrors the structure of the classical theory of concepts, and therefore it falls under the strokes of Cassirer's criticisms. Still, it is worth reading this logic textbook to better comprehend the difference between a 'really Kantian approach' to the theory of concepts (that is, the critic-transcendental one) and a 'classical logical approach taught by Kant' to his students. In this way, not only can the difference between the two doctrines be emphasized again, but it also becomes possible to tackle notions never discussed in the *Critique*. For example, only in the *Logic* does Kant face the question of what the notion of 'characteristic' is; a question that we will find again with Frege.

We can start the survey of the *Logic* by recalling that: 'human knowledge [...] is discursive, that is, it takes place through representations that make what is common to several things the ground of knowledge, thus through characteristics as such' (Ibid., pp. 63–4). Kant refers to the characteristics, and gives this definition: 'a characteristic is that in a thing which makes up part of its knowledge, or – which is the same – a partial representation so far as it is concerned as cognitive ground of the whole representation' (Ibid., p. 64).

Let us notice that Kant speaks of characteristics as parts of the conceptual knowledge of an object. With reference to this aspect, Kant differentiates between 'concept' and 'object', or, rather, between the '*form* of the concept' (its universality) and the '*matter* of the concept' (the object) (Ibid., p. 96). Moreover he also distinguishes 'attribute' from 'characteristic': the former is a 'necessary derivative characteristic', that is, the characteristic which must always exist in the represented thing, even if it is derived from other necessary characteristics. Having 'three angles' is an attribute of triangles, since it is a characteristic derived from the necessary characteristic of having 'three sides' (Ibid., pp. 66–7). Note that even if Kant speaks about necessary characteristics, he intends an *epistemological necessity*. Therefore, when he considers the essence of a thing, he is actually referring to the set of predicates that are needed to individuate it as that thing. The *epistemological essence* has nothing to do with the *real essence* (Ibid.): the metaphysical one.

Kant's theory of characteristic has to be completed with the observation that when a characteristic is conceived as a single unit, it is a representation, a concept. If it is considered in relation to knowledge of the object as a whole a characteristic is instead a partial representation (*Teilvorstellung, Partialvorstellung, Teilbegrif*) of the concept under which the object falls. From this point of view, Kant can claim that the more specific a concept is, the more characteristics it has to be made up of. Those characteristics are parts of the concept, and allow us to know the object to which it (the concept) refers. For example, the concept 'ball' is made up of the partial concepts (characteristics) 'solid', 'elastic', 'heavy', and so on.

There are, thus, two uses of the term 'characteristic': (1) what is shared by different concepts of different objects; and (2) what makes up part of the representation of a thing, and which, then, allows us either to know a given thing (internal use of characteristics), or to compare two things according to their similarities and their differences (external use of characteristics) (Ibid., p. 64). Among the several distinctions concerning characteristics, the one between *analytical characteristics* and *synthetic characteristics* is particularly important. The former are those partial concepts necessarily belonging to the concept *qua* concept; for example, in the case of 'red ball', the analytical characteristics are 'ball', 'red', 'elastic', 'solid' and so on. The latter are those partial concepts contingently belonging to the concept; for example, in the case of 'red ball', synthetic characteristics can be 'odorous', 'smooth' and so on. The same distinction can be traced from a different point of view, that is, that one concerning the ways in which characteristics are correlated: there are subordinate characteristics (in a subordination relation) and coordinate characteristics. The former are those that are set one *under* another, and so they make up a *series*. The latter are those that can be set one *after* another and make up an *aggregate* (Ibid., p. 65). It follows that:

> every concept, as a partial concept, is contained in the representation of thing; as a ground cognition, that is, as a characteristic, it has these things contained under it. In the former regard, every concept has an intension [content]; in the latter, it has an extension. Intension and extension of a concept have an inverse relation to each other. The more the concept contains under it, the less it contains in it.
>
> (Ibid., pp. 101–2)[17]

Therefore, each concept can be considered extensionally (that is, as the set of objects falling under it), or as compound of characteristics (that is, as having a content, a comprehension). Consequently, the bigger its extension is, the smaller its content is: 'the concept of *body*, for example, has a greater extension than the concept of *metal*' (Ibid., p. 102), but it has a smaller content of characteristics.

Let us note that this is exactly one of the consequences of the theory of concepts rejected by Cassirer. According to the Marburgh philosopher, a suitable doctrine of concepts must allow us to speak both of greater extension and greater comprehension. Cassirer found such a doctrine of concepts in Kant; but, of course, not in the Kant of *Logic*, where the classical theory still has a great influence, but in the Kant of *Critique*: the 'real' Kant. In fact, while discussing the Kantian innovations concerning the doctrine of concepts, Cassirer wrote:

> now it no longer suffices to take it as a mere generic concept, a conceptus communis. For such a concept is lacking in precisely the characteristic and decisive factor; it is a mere expression of the analytical, but not of synthetic unity of consciousness. But it is only the

> previous thought of a synthetic unity that makes an analytical unity thinkable.
>
> (Cassirer, 1923–29, Vol. 3, p. 316)

Let us come back to the concepts of 'body' and 'metal'. Kant observes that there is a relation between higher concepts and lower concepts grounded in the fact that (in accordance with Aristotle's principle of predication) 'a characteristic of a characteristic – a distant characteristic – is a higher concept; the concept in reference to a distant characteristic is a lower concept' (L, p. 103).

In this regard, Kant makes a distinction (which we will also find in Frege):

> the lower concept is not contained in the higher, for it contains more in itself than the higher; but it is yet contained under the latter, because the higher contains the cognitive ground of the lower. Further, a concept is not wider than another because it contains more under it – for one cannot know that – but so far as it contains under it the other concept and besides it still more.
>
> (Ibid., p. 104)

So, the concept 'ball' is not contained *in* the concept 'body', since 'ball' has more characteristics than 'body'; but the concept 'ball' is contained *under* the concept 'body'.

Finally, in the relation between higher and lower concepts, Kant proposes something similar to the Porphyry-tree and, consequently, he speaks of highest genus and lowest species as well. Traditionally, he defines the highest genus as that genus which cannot be a species, and the lowest species as the species which cannot be a genus. Naturally, each concept which is neither highest genus, nor lowest species can be both a genus (as far as the subsumed species are concerned), and a species (as far as the genera under which it is subsumed are concerned). The only aspect of the Porphyryan structure that Kant does not accept is that it has an end, at its bottom. According to Kant, even if we could reach a concept which could be applied directly to an individual thing, we could always add some further logical determinations to it, and so a new concept subsumed under it would be found. And so on. Had we in fact the complete totality of logical determinations, we would have an intuition (not a concept) (Ibid., pp. 103–4), that is, something which cannot be an object of knowledge. The only possible object of knowledge is a categorized intuition, and this object is something which we have already delimited by means of our conceptual structure. In conclusion, Kant does not close the bottom of the Porphyry-tree with individuals, as Aristotle and the medievals did – even if he takes this possibility into account – but only for pragmatic purposes and as a temporary assumption.

1.1.3 Cassirer: from class-concept to function-concept

According to Cassirer, it is mathematics that points out the problems of the class-concept theory. But it is also mathematics that tells us how to overcome them. First of all, Cassirer completely rejects the classical thesis of concept genesis, which was grounded in the process of abstraction, in the sense both of *abstrahere aliquid* and of *abstrahere ab aliquo*. Concept has to be the result of a productive act of thought, as happens in the mathematical generalization that consists in searching for a new expression from which the starting expression can be derived. This is not a passage from a less general and more determined concept (as far as its characteristics are concerned) to a more general but less determined concept. Actually, it is the production of a perfectly determined concept from which derived concepts can be drawn as particular cases:

> The genuine concept does not disregard the peculiarities and particularities which it holds under it, but seeks to show the necessity of the occurrence and connection of just these peculiarities. What it gives is a universal rule for the connection of the particulars themselves. Thus we can proceed from a general mathematical formula, – for example, from the formula of a curve of the second order, – to the special geometric forms of the circle, the ellipse, and so on, by considering a certain parameter which occurs in them and permitting it to vary through a continuous series of magnitudes.
>
> (Cassirer, 1910, pp. 19–20)

Thus, mathematical thinking transforms the world by means of 'forms of relation', that is, by means of 'an ordered system of strictly differentiated intellectual *functions*' (Ibid., p. 14), due to a creative (productive) act of the knowing subject.

The concept reached thus is not only more universal than the concepts which can be derived from it, but it has also a larger comprehension. In this way one problem of the classical theory can be avoided, that is, that one arising from the fact that the larger a concept's universality, the smaller its comprehension.

The concept should no longer be understood as a class containing objects, but as a representation unifying given objects in a given way by giving them cognitive significance. Consequently, *the form of a concept* has a fundamental and primary role: to rule the intellectual synthesis, by means of which we constitute as a 'series' (that is, in Cassirer's words, as a set ordered from a given point of view) what otherwise would not be a 'series'.

We are dealing with a clearly Kantian theoretical perspective: knowledge is marked by normativity, by the priority of representations in respect to objects. And the concept is the unity of representation. It is the form by means of which something can be understood as an object (*the content of the concept*) belonging to a given series, that is, to a given class of objects. In this way, Kant's 'Copernican Revolution' is completely taken over:

> The concepts of the manifold species and genera are supposed to arise for us by the gradual predominance of the similarities of things over their differences, that is, the similarities alone, by virtue of their many appearances, imprint themselves upon the mind, while the individual differences, which change from case to case, fail to attain like fixity and permanence. The similarity of things, however, can manifestly only be effective and fruitful, if it is understood and judged as such [...] This transition from member to member, however, manifestly presupposes a principle according to which it takes place, and by which the form of dependence between each member and the succeeding one, is determined. Thus from this point of view also it appears that all construction of concepts is connected with some definite form of construction of series.
>
> (Ibid., p. 15)

The classical thesis according to which it is sufficient to observe objects in order to find out similarities and differences is wrong: *similarities are not in objects, but in the ways in which the knowing subject cognitively looks at them*. It is the knowing subject who groups the objects in a 'series', since he or she possesses the concept of that series. How could we claim that those objects have the same property, had we not the concept correlated with that property? What, if not the correlated concepts, would allow us to render them as cognitively significant? Therefore, this synthetic moment must be epistemologically first, that is, it should be considered as an a priori moment. But, note, it is a priori only in the epistemological sense: it must precede our knowledge of the world since it allows us such knowledge. Concepts are not the results of inductive abstractive processes; conversely, they are the starting points of synthetic acts by means of which the world is ordered, that is, rendered cognitively significant. The concept is not the container defined by the objects it contains; rather it is the shape of the container, in which only objects of the right shape can be collocated. This Kantian background notwithstanding, in *Substance and Function*, Kant is never named. Nevertheless, Cassirer pays his debt openly in *The Philosophy of Symbolic Forms*, where he also remarks that:

> the theoretical concept in the strict sense of the word does not content itself with surveying the world of objects and simply reflecting its order. Here the comprehension, the 'synopsis' of the manifold is not simply imposed upon thought by objects, but must be created by independent activities of thought, in accordance with its own norms and criteria.
>
> (Cassirer, 1923–29, Vol. 3, p. 284)

This is exactly Kant's idea of concept as 'unity of rules'. However, Cassirer goes further than Kant, and tries to mix the Kantian idea of function with the idea of function that was being developed in works on the foundations of mathematics in those years – in Russell's *Principles of Mathematics* of 1903, in Russell's and Whitehead's *Principia Mathematica* of 1910, and in the earlier essays by Frege, *The Foundations of Arithmetic* (1884), and 'A Critical Elucidation of some Points in Schröder's *Vorlesungen über die Algebra der Logik*' (1895).

> There are, he [Russell in *Principles of Mathematics*] stresses, two ways to determine classes: one by pointing out their members one by one and connecting them as a mere aggregate, by a simple 'and' – the other by stating a universal characteristic, a condition which all members of the class must fulfill. Russell sets this latter generation of the class, the 'intensional', over against the former, which is explained by means of 'extension'. And they do not remain in such juxtaposition, for it becomes gradually clearer that the definition by intension has precedence over the definition by extension.
>
> (Cassirer, 1923–9, Vol. 3, p. 294)

This is a different way of explaining the passage from class-concept to function-concept. It is the passage from extensionality to that form of intensionality connected with the rule, with the function. Furthermore, by using a terminology taken – as Cassirer explicitly admits – from the *Principia Mathematica*, he notices that the elements of a class have to be thought of as 'variables of a determinate propositional function' (Ibid., p. 295). Here,

> the propositional function as such must be strictly distinguished from any particular proposition, from a judgment in the usual logical sense. For what it primarily gives us is only a pattern for judgments but in itself is no judgment: it lacks the decisive characteristic of judgment, since it is neither true nor false. Truth and falsity attach only to the individual judgment in which a definite predicate is related to a definite subject; whereas the propositional function contains no such definiteness but only sets up a general schema which must be filled with definite values before it can achieve the character of a particular proposition [...] In this sense, every mathematical equation is an example of such a propositional function. Let us take the equation $x^2-2x-8=0$. This expression is true if for the (initially wholly undetermined) values of x we substitute the two roots of the equation; for all other values, it is false. On the basis of these determinations, we can give a genuine purely 'intensional' definition of class. If we consider all *x*'s so constituted that they belong to the type of a certain propositional function $\phi(x)$ and group together the values of *x* which prove to be 'true' values for this function, we have defined a determinate class by means of the function $\phi(x)$ [...] But with this, what logic calls a concept has by no means been broken down into a collective quantity; on the contrary, the quality is once again grounded in the concept.
>
> (Ibid., pp. 295–6)

This is a decisive point. *The concept is a propositional function containing two elements having a completely different nature*:

> the universal form of the function as designed by the letter ϕ stands out sharply against the values of the variable *x* which can enter into this function as true variables. The function determines the relation between these values, but it not itself one of them: the ϕ of *x* is not homogeneous with *x* the series, x_1, x_2, x_3 and so on.
>
> (Ibid., p. 301)

Thus, according to Cassirer there is a sharp distinction between the concept as form and its content, that is, between *concept and object*:

> In this conception there is no danger of hypostatising the pure concept, of giving it an independent reality along with the particular things. The serial form *F*(*a*, *b*, *c*,...) which connects the members of a manifold obviously cannot be thought after the fashion of an individual *a* or *b* or *c*, without thereby losing its peculiar character. Its 'being' consists exclusively in the logical determination by which it is clearly differentiated from other possible serial forms φ, ψ, ...; and this determination can only be expressed by a synthetic act of definition, and not by a simple sensuous intuition.
>
> (Cassirer, 1910, p. 26)

> [Thus] if we wish to conceive of the concept itself, we must not attempt to clutch it like an object.
>
> (Cassirer, 1923–29, Vol. 3, p. 299; compare also pp. 315–16)

Therefore, by accepting that the nature of the concept is formal and functional and that the concept is distinct from the object, Cassirer reaches two results: (1) he dissolves the problem of universals (that is, the concept can be seen neither as a real entity, nor as a *flatus vocis*); (2) he dissolves the classical dichotomy particular–universal.

With reference to the particular intuition, which is always *hic et nunc*, Cassirer clearly shows that it cannot be the starting-point for the abstraction leading to the concept (since the latter is an intellectual product), but it must be thought of as a catalyst (Ibid., Part III, Ch. V, § 1). From this point of view the particular intuition is the first (even if contingent) cause of either the application of concepts which are already possessed, or the formation of new ones. In conclusion, the role of empirical intuition in Cassirer's view is completely Kantian, and Cassirer would have agreed with Vaihinger – himself an interpreter and follower of Kant – when he wrote:

> Just as *Meleagrina margaritifera*, when a grain of sand gets beneath its shining surface, covers it over with a self-produced mass of mother-of-pearl, in order to change the insignificant grain into a brilliant pearl, so, only still more delicately, the psyche, when stimulated, transforms the material of sensation which it absorbs into shining pearls of thought, into structures.
>
> (Vaihinger, 1911, p. 7)

Empirical intuition is cognitively significant only '*sub specie postulati*'. In other words, the empirical fact is always theoretically oriented: 'what is here called the "object" of knowledge acquires determinate meaning only by being referred to a certain form or function of knowledge' (Cassirer, 1923–29, Vol. 3, Part II, p. 321; and passim).

This amounts to saying that empirical facts are always theory-laden. It follows that the universal and the particular share the same nature: as soon as the particular becomes meaningful for the knowing subject, the latter also grasps the universal which makes that particular cognitively significant, which constitutes it. As soon as something is identified as the member of a 'series', immediately the 'series' (that is, the rule, the form of the concept under which that element falls) is grasped. So, even if there is a difference between universal and particular, between form and content, between concept and object, that difference is 'not in being but in meaning' (Ibid., Vol. 3, Part III, Ch. I).

At this point it should be clear that, though we produce concepts by means of intellectual acts, *concepts do not construct objects*, but, à la Kant, *concepts cognitively constitute objects*. Furthermore, the production of the concept is hypothetical:

> A totality of members *a, b, c, d*... which are at first given solely in their 'thatness', in the actuality of their spatio-temporal togetherness, are to be recognized as belonging together, are to be linked by a rule on the basis of which the production of the one from the other can be determined and foreseen. This law of production is never immediately given in the same way as perceptions: it must be injected into them in a purely intellectual, hypothetical way.
>
> (Ibid., Vol. 3, Part III, Ch. V, § 1)

> The conceptual system does not have the truth and certainty which can be guaranteed by the truth and certainty of the framework of categories, as it did in Kant. Now the framework of categories has to be constructed in a hypothetical way. Thus, the knowledge of the world cannot be determined once for all by something like the 'the laws of the intellect', but varies as soon as we change our conceptual framework, or better the form of that framework: 'every transformation of the genuinely "formal" concept produces a new interpretation of the whole field that is characterized and ordered by it'.
>
> (Cassirer, 1910, p. 26)

In conclusion: (1) the concept is a knowing subject's hypothetical production; (2) it allows us the constitution of the cognitive significance of the objects falling under it, therefore it is productive, rather than re-productive; (3) it is a projection of the cognitive significance, that is, it sets up a point of view on possible objects which could fall under it, even if without any commitment on their existence (Cassirer, 1923–29, Vol. 3, Part III, Ch. I).

On functions

From what has been said, it appears that in Cassirer's analysis there are two notions of 'function'. The first is function à la Kant, that is, function as the 'unity of the act which orders several representations under a common representation'. The second is function à la Frege; that is, function in the mathematical sense.

About the former, we should note that, whereas in *Substance and Function*, Cassirer attempts to find a Kantian foundation only for the natural sciences, in *The Philosophy of Symbolic Forms*, he tries both to enlarge his purpose and to give a unitary foundation to all cultural forms. This attempt is not carried out on the ground of a substantial-metaphysical unity, but on the grounds of a functional unity, that is, on the ground of the fact that knowledge is always *knowledge by means of functions*, in any cultural context (science, myth, art, religion and so on). In other words, in *The Philosophy of Symbolic Forms*, Cassirer's aims are to shift the cognitive analysis from objects to functions, and to explain any possible cultural form as a system of symbols, which are understood as material manifestations of the respective functions. Therefore, any intellectual function is symbolic; this is why Cassirer (1944) defines 'man' as an *animal symbolicum*.

Into this general philosophical framework, Cassirer inserts the interpretation of conceptual function as mathematical function. But, does it fit well? Is it correct? Can one really mix the Kantian and the Fregean interpretations together, as Cassirer would like? Certainly, every mathematical function can be seen as an act of synthesis. But can every act of synthesis be interpreted as a mathematical function? In order to answer these questions, we should first examine the origin of this conception, that is, we must consider Frege.

1.1.4 Frege: from the concept of function to the function-concept

'What is a function?' is the emblematic title of the essay written by Frege for L. Boltzmann's *Festschrift* in 1904. Yet, since 1879, the year in which he published his *Conceptual Notation*, Frege had been dealing with this central question for his innovations in both logic and theory of concepts. To begin with this question, let us distinguish between function and its graphical expression, that is, its sign. Thus let us recall the well-known distinction between *Sinn* and *Bedeutung*,[18] introduced by Frege in 1892. Consider the function expressed by the sign $f(x)=2x^3+3$. In this case, the function (the *Bedeutung* of this sign) maps the number (the *Bedeutung*) referred, for example, to the numeral 0 (the sign) into the number (the *Bedeutung*) referred to the numeral 3 (the sign); the number referred to the numeral 1 into the number referred to the numeral 5; the number referred to the numeral 2 into the number referred to the numeral 19; and so on. Therefore, on the one hand, the function is the *Bedeutung* of a linguistic expression (in this particular case, a function expression), and, on the other hand, it is an operator mapping numerical values into numerical values, namely, in Frege's words, objects into objects. However, its main feature concerns its being something which 'must be called incomplete, in need of supplementation, or "unsaturated"' (Frege, 1891, p. 140).

It should be noted that *it is the function which is incomplete, not its linguistic expression*. Nevertheless, to make this incompleteness graphically clear, the sign referring to such a function should be written with blanks in it:

$$f(\ )=2(\ )^3+3.$$

The argument (an object) completes, or saturates, the function. Therefore the sign referring to such an argument completes the expression referring to the function.

Thus, there are two completely heterogeneous parts: the function, which needs to be saturated, and the argument saturating it. In the case of mathematics, after we have filled a function with an argument (a numerical value), it allows us to find another numerical value; that is, the value of the function for that particular argument. We have, then, a set of ordered pairs: the first element is one of the possible values of the argument (they make up the *graph* of the function); the second element is the corresponding value of the function.

Besides this notion of function, which I call *type-I-function*, Frege proposes another notion, which I call *type-II-function*, that is essential to grasp his theory of propositions. Type-II-function is the result of an enlargement of the notion of function by means of an enlargement of the 'formation of a function'; that is, of the way of constructing it:

> I begin to add to the signs +, −, and so on, which serve for constructing a functional expression, also signs such as =, >, <, so

that I can speak, for example, of the function $x^2 = 1$, where x takes the place of the argument, as before.

(Ibid., p. 28)

Usually, $x^2 = 1$ is not thought of as a function, but as an equation, that is, as an expression which states an identity between two type-I-functions:[19] the function $f(x) = x^2$, and the function $g(x) = 1$. Therefore, beyond the type-I-function, we also have the type-II-function, that is, functions such as $(x^3 + 2 < 3x)$, or $(x^2 - 1 = x - 1)$.

Nevertheless a type-I-function is neither true nor false, it can only admit some values for some arguments belonging to its field of existence. An equation however, can be true (correct) for some arguments and false (incorrect) for others. Thus, only a type-II-function can be true or false. We know that a type-I-function is an operator that maps numbers (the values of the argument) into numbers (the values of the function), that is, objects into objects. Also in the case of type-II-functions (for example, $x^2 - 1 = x - 1$) the objects mapped are numbers. However, the objects into which they are mapped (the images of the former, in modern terms) are, for Frege, the truth-values 'True' and 'False', according to whether they are saturated with the right or the wrong argument. In summary, we have: (1) type-I-functions, in which there are no signs such as $=$, $<$, $>$; they map numbers into numbers; (2) type-II-functions, in which there are signs such as $=, <, >, \ldots$; they are relations among type-I-functions; they map numbers into truth-values. Therefore, we must pay attention in writing $x^2 - 4x = x(x - 4)$, which can be interpreted either as a type-II-function, that is, a function which maps numbers into truth-values, or as an identity between two type-I-functions: $f_1 = x^2 - 4x$, and $f_2 = x(x - 4)$.[20]

Under this formulation there is a new theory of proposition based on the distinction between function and argument. Rather, we should introduce the distinction between the sign and what it stands for, and therefore we should claim that a proposition is made up of two parts: the sign of the *argument* – the singular term – and the sign of the function – the *predicate*, or *conceptual term*, or *conceptual word*, as Frege calls it. These two, then, refer, respectively, to a saturating object and to a function which needs to be saturated. If the completion is correct, that is, if the argument is the right one, a true proposition is obtained ('Socrates is a man'); if the completion is wrong, a false proposition is obtained ('Socrates is a dog').

It follows that there is a structural analogy between type-II-functions like $x^2 - 1 = x - 1$ and expressions like 'x is a man'. The former is an equation (type-II-function), which can be true or false, depending on what the saturating arguments are. Similarly, the latter is a linguistic expression, which can be true or false, depending on whether it is, or it is not, correctly saturated. Thus: 'the linguistic form of equations [interpreted as type-II-functions] is a statement' (Ibid., p. 146).

Consequently, as function and argument are two completely distinct and heterogeneous things, so the concept to be saturated and the saturating object must be completely distinct and heterogeneous. Similarly, the incomplete conceptual term denoting the concept and the saturating singular term denoting the object are also completely distinct and heterogeneous. Furthermore, if we bear in mind the distinction between type-I-functions and type-II-functions and, thus, between their signs, we can easily understand that every conceptual term can be thought of as a functional expression, whereas not all functional expressions can be thought of as conceptual terms. This happens only for the functional expressions standing for type-II-functions, that is, only for those containing signs such as =, <, >, and so on. Only these can be true and false, not those standing for type-I-functions.

Within this framework of similarity between functional expression and conceptual term, Frege identifies a concept as the *Bedeutung* of the latter. Therefore he considers it as something completely different from the universal associated to a predicate (or to the subject), as it was in the classical analysis of proposition. According to this new interpretation, concepts are unsaturated parts, and, thus, they are type-II-functions (Ibid., p. 30).

To sum up, the concept (the *Bedeutung* of a conceptual term) is a type-II-function that maps objects (the *Bedeutungen* of singular terms, that is, the saturating objects) into objects, that is, the truth-values (the *Bedeutungen*) of the saturated propositions.

But, what is *Sinn*? In the case of a singular term, the *Sinn* is what it objectively expresses, and what allows its *Bedeutung* to be grasped. Unfortunately, in the case of an unsaturated expression (a conceptual term) to grasp what the *Sinn* is is more difficult. The problem lies in the fact that Frege does not say anything about this. Nevertheless, he explicitly writes that there is a *Sinn*.[21] In this way, he leaves an unresolved question, and uncountable attempts have been made to solve it. Consequently, there is great disagreement among those who have tried to complete Frege's framework, and thus to make Frege say what he did not say, but what they think he would have said.

With reference to the *Sinn* of a proposition, Frege writes several clear pages to suggest that it is the *thought* expressed by the proposition.

That is, it is something objective, belonging to a 'third realm' having strong Platonic features, and therefore quite different from the subjective *image*[22] that we make of an object. Considering our aims, it is enough to point out that, according to Frege, knowledge is strictly connected to the relation between the thought and the truth-value of the proposition expressing the former, and that there is a difference between *thinking* and *judging*. Thinking concerns the grasping of the *Sinn* of a proposition, that is, of the thought it expresses. Judging concerns the recognition of the thought as a true thought: 'judgments can be regarded as advances from a thought to a truth-value' (Frege, 1892a, p. 65).

Thus every proposition expresses a thought. But Frege is interested only in the true thoughts. This is crucial to correctly understanding his proposal. Only the true thoughts provide knowledge, in particular, scientific knowledge.

Up to this point, I have suggested an interpretation starting from the notion of 'function', as incomplete, unsaturated, needing a completion element, and from the notion of 'argument', as a complete and saturating element. Then, I have shown how the concept can be interpreted as a type-II-function, and, thus, how it is incomplete, unsaturated, and therefore needing a completion. Lastly, I have stressed that function and concept are the *Bedeutungen* of linguistic signs (the functional expressions). When the latter have been saturated, they give rise to propositions, whose *Bedeutung* is a truth-value and whose *Sinn* is a thought.

This can be thought of as a latent argument supporting a particular interpretation of the Fregean notion of 'concept'. Let us consider this point. There are two opposite views. According to the first, what is unsaturated is primarily the concept, and as a consequence, since the concept is the *Bedeutung* of a linguistic expression, this latter is also unsaturated. According to the second, what is unsaturated is primarily the linguistic expression, and as a consequence, since the concept is its *Bedeutung*, this latter is also unsaturated. As should be clear, I support the first view, and, thus, I suggest that, according to Frege, the thought comes before the language; whereas the contrary should be considered as a misinterpretation of Frege's conception.

As in the case of Aristotle's doctrine of categories where there was a debate between the supporters of the logical approach, the linguistic approach and the ontological approach, in the case of Frege's doctrine there are supporters of the linguistic approach, especially Dummett (1973), and supporters of an ontological approach, mainly, but not completely, represented by Sluga (1980). In particular, Dummett suggests that Frege's conception represents a drastic and revolutionary turn in the development of western philosophy, something independent of what happened before. According to Sluga, though, Frege is firmly anchored to his German cultural background and his proposal is not completely revolutionary. Besides Sluga other scholars also argue that the thesis of such independence cannot be supported. For example, Gabriel (1986) shows that Frege was in touch with the Kantian tradition, that is, that he was linked to the coeval German philosophical mainstream. Is such a link merely terminological, as some supporters of his independence claim? I do not think so. For example, rather than from functions and concepts, Frege moves from propositions and he speaks of concepts as predicates of possible judgements, exactly as Kant did: 'As opposed to this [Boole's conception], I start out from judgments and their contents, and not from concepts [...] I only allow the formation of concepts to proceed from judgments' (Frege, 1880–81, p. 16). This is because his conceptual notation, quoting Leibniz, should '*peindre non pas les paroles, mais les pensées*' (Ibid., p. 13).

If the ground of the analysis was the proposition, there would be good reason to argue for the priority of language. Although, the antecedent is not right if one thinks of the true propositions as signs of the true thoughts: the real foundation. With reference to this point, it should be recalled that in *Categoriae* and in *De Interpretatione*, Aristotle himself started off from a linguistic element – exactly like Kant and Frege – but he warned the reader that what is first for us (language) is not first for nature. Similarly for Frege, the thought (especially, the true thought) must be considered 'first for nature', to use an Aristotelian phrase, and, therefore the *Bedeutungen* of the singular terms correctly saturating the propositions have a particular importance. Thus, Aristotle, Kant and Frege move from propositions. But according to Aristotle and Frege propositions are the linguistic basis from which to start discovering the real foundation (in Aristotle, the being and its logical structure; in Frege, the thought and its logical structure); whereas according to Kant they are the transcendental starting point. At least from this point of view, Frege is more Aristotelian than Kantian, even if his concern for a realm of objective thoughts is quite Platonic. We could claim, at least in relation to this aspect, that Frege's method is Aristotelian but that his aim (that is, the construction of a perfect language mirroring the realm of the objective thoughts) is Platonic. Note that the secondariness of linguistic expressions with respect to thought is explicitly expressed in a passage of the *Posthumous Writings*: 'we shall not derive thinking from speaking; thinking will then emerge as that which has priority and we shall not be able to blame thinking for the logical defects we have noted in language' (Frege, 1924–25, p. 270).

This interpretation, according to which there is *a priority* of thought over language, has two other strong supporting points: (1) Frege often criticizes what is first for us, that is, the natural language; (2) he attempts to construct a perfect language which can faithfully mirror the realm of the thought:

> what I am striving after is a *lingua characteristica* in the first instance for mathematics, not a *calculus* restricted to pure logic. But the content has to be rendered more exactly than is done by verbal language. For that leaves a great deal to guesswork, even if only of the most elementary kind.
>
> (Frege, 1880–81, p. 12)

Certainly Frege writes a paper on the philosophy of language, that is, 'Über Sinn und Bedeutung'. However, this should be considered as something functional for a project which, among other things, also requires the distinction between two elements such as *Sinn* and *Bedeutung*. It should not be seen as the unmistakable mark of an ontological preference for language. That this paper can be read as the starting point for a new way of doing philosophy is a different kettle of fish altogether, but – I suspect – the kettle is not Frege's and neither is the fish. One could object that Frege did start from an analysis of natural language and enlighten its ambiguities, and that he went on to emendate them in a perfect language. Yet, to think of a perfect language he must have had in his mind a criterion for a perfect language, and this criterion can be found in his doctrine of true thought. Then, in this case too the thought would be prior.

Frege criticizes natural language, which can draw us to the misguided distinction between subject-predicate, and he makes a plea for a perfect language, which is isomorphic to the thought and where that ambiguity does not exist. Moreover several times Frege writes that natural language is ambiguous, since it can disguise thought in very different ways, as, for example, happens in the case of active and passive forms:

> [Let us consider] the two propositions 'the Greeks defeated the Persians at Platea' and 'the Persians were defeated by the Greeks at Platea' [...] Now I call the part of the content that is the same in both the *conceptual content*. *Only this* has significance for our symbolic

language; we need therefore make no distinction between propositions that have the same conceptual content.

(Frege, 1879, p. 3)

What is important is the thought and how it is obtained by correctly saturating a concept with an object. Is this not a priority of thought over language?[23]

Objects, concepts, properties

We have met the term 'object' several times, and we have always considered it as the *Bedeutung* of something, *in primis* the *Bedeutung* of singular terms. In more general terms, 'object' is something completed. Besides this Frege does not compromise himself, since he thinks a 'scholastic definition' of object cannot be given, being 'something too simple to admit of logical analysis. It is only possible to indicate what is meant. Here I can only say briefly: an object is anything that is not a function, so that an expression for it does not contain any empty place' (Frege, 1891, p. 32). Nevertheless, even if we do not have a precise definition, we have some fragments indicating what can be considered as an object. There is a fundamental ambiguity, though, and several interpreters disagree about Fregean ontology (see Klemke, 1968). Certainly the *Bedeutungen* of singular terms are objects. Truth-values are objects (Frege, 1892a, p. 63). The graphs of functions are objects (Frege, 1891, p. 32), and thus, as we shall see, the extensions of concepts are objects (Frege, 1892–95). Can anything else be considered as an object?

Frege does not want to give a 'true and proper' definition of concept either, as 'one cannot require that everything shall be defined, any more than one can require that a chemist shall decompose every substance' (Frege, 1892b, p. 42). However, the ambiguity about concepts is less, since a 'concept (as I understand the word) is predicative... (fn: It is, in fact, the *Bedeutung* of a grammatical predicate)' (Ibid., p. 43). Furthermore, a concept is a type-II-function (Frege, 1892–95).

Concept and object are two completely different things, exactly like function and argument are. This notwithstanding, there are some relations between them. First of all, objects *fall under* concepts. The object 'Socrates' falls under the concept 'to be a man', in the sense that the former correctly saturates the concept '*x* is a man' and, therefore, it makes the proposition 'Socrates is a man' true. However, there are not only relations between objects and concepts, but relations between concepts as well. There are quantified propositions, such as 'All men are mortal', 'There is a white cat', where we are dealing with the relation between first level and second level concepts, that is, with the fact that a first level concept *falls within* a second level concept. To understand this idea, Frege suggests reflecting on the negation of a quantified proposition:

> In universal and particular affirmative and negative sentences, we are expressing relations between concepts; we use these words ['all', 'any', 'no', 'some'] to indicate the special kind of relation. They are thus, logically speaking, not to be more closely associated with the concept-words that follow them, but they are to be related to the sentence as a whole. It is easy to see this in the case of negation. If in the sentence 'all mammals are land-dwellers' the phrase 'all mammals' expressed the logical subject of the predicate 'are land-dwellers', then in order to negate the whole sentence we should have to negate the predicate: 'are not land-dwellers'. Instead, we must put the 'not' in front of 'all'; from which it follows that 'all' logically belong to the predicate.
>
> (Frege, 1892b, p. 48)

Therefore, the quantified proposition 'Everybody is alive' is true granted that (1) the second level concept related to the quantifier 'Everybody' is saturated by a first level concept 'x is alive', and (2) this latter, in its turn, is truly saturated by all objects in the domain. Similarly, the quantified proposition 'Something is a flower' is true granted that (1) the second level concept connected to the quantifier 'Something' is saturated by a first level concept ('x is a flower'), and (2) this is truly saturated by at least one object in the domain.

Thus, whereas an object can *fall under* a concept, a first level concept can *fall within* a second level concept. Moreover, besides the relation between first and second level concepts, there is another relation, that is, the one of *subordination* among first level concepts.

This relation is grounded on the notions of 'property of an object' and 'characteristic of a concept'.[24] As far as this question is concerned, Frege states that *something can be a property and a characteristic at the same time, even if it cannot be such with reference to the same thing: properties are concerned with objects, characteristics are concerned with concepts*

Let Γ be an object and Φ be a concept, then we can claim that

Φ is a *property* of Γ,

or that

Γ *falls under* the concept Φ.

Let us now suppose that the same object Γ *has the properties* Φ, X, Ψ, which we group together in the property Ω. Φ, X, Ψ *are the characteristics of the concept* Ω. That is,

Φ, X, Ψ are *properties* of Γ,

or

Ω is a *property* of Γ,

or

Γ *falls under* the concept Ω,

or

Φ, X, Ψ are *characteristics* of the concept Ω.

Φ, X, Ψ are properties of the object Γ, but also characteristics of the concept Ω. Thus, the property of an object is a concept, in particular, that under which the object falls. But also the characteristic of a concept is a concept, in particular, that concept that subordinates the concept of which it is a characteristic. Ω, in the example above, is subordinate to the concept Φ, to the concept X and to the concept Ψ. If the object is the number 2, and the properties are: 'to be a positive number', 'to be a whole number', 'to be divisible by two', therefore the object 2 falls under the concepts 'to be a positive number', 'to be a whole number', 'to be divisible by two', that is, under the concept 'to be a positive, whole number, divisible by two'. The characteristics of the latter concept are exactly those concepts which are the properties of the object 2, and, thus, it – the concept at issue – is subordinate to them.

Frege emphasizes that the subordination relation (between characteristics and concepts) is different from both that between objects and properties (*falling-under* relation), and that between first and second level concepts (*falling-within* relation): 'a concept can fall under a higher concept [... this relation] however, must not be confused with one concept's being subordinate to another' (Ibid., p. 45). In other words:

> Γ falls under the concept Φ; but Ω, which is itself a concept, cannot fall under the first-level concept Φ; only to a second-level concept could it stand in a similar relation. Ω is, on the other hand, subordinate to Φ [... For example, let us consider the concept] *positive whole number less than 10*. This is neither positive, nor a whole number, nor

less than 10. It is indeed subordinate to the concept *whole number*, but does not fall under it.

(Ibid., pp. 51–2)

Therefore, in the proposition 'Fido the dog is a mammal', the fact of being a mammal is a property of the object denoted by the singular term 'Fido the dog', and thus that object *falls under the concept* denoted by the conceptual term 'being a mammal'. Nevertheless, note that the case of the proposition 'All whole positive numbers smaller than ten are positive numbers' is slightly different. Here there are both the *falling-within* relation and the *subordination* relation. For it is a quantified proposition, and thus there are first level concepts ('being a whole, positive number smaller than 10' and 'being a positive number') falling within the second level concept which is related to the quantifier 'All'. Furthermore, the (first level) concept 'being a whole, positive number smaller than 10' is subordinate to the (first level) concept 'being a positive number', since the latter is a characteristic of the former.

Consequently, in any non-quantified proposition there are only objects falling under concepts; whereas in quantified propositions there are concepts falling within higher order concepts. Furthermore, there can be first level concepts subordinate to other first level concepts. However, even if Frege introduces the fundamental distinction between object and concept, the distinction between property and characteristic is still ambiguous, since they are both the same concept. If we pay attention, we can realize that this distinction (and so the distinction between 'falling under' and 'subordination') recalls a distinction we have already considered in Kant's *Logic*. Frege speaks of subordination between two first level concepts when one is a characteristic of the other, and he distinguishes it from both the falling-within relation and the falling-under relation. Kant spoke of the subordination relation between lower concept and higher concept, where the former does not 'fall in', but 'falls under' the latter. Frege affirms that some concepts are properties in relation to the objects which fall under them, and, at the same time, they are characteristics in relation to the concepts which are subordinate to them. Kant affirmed that all concepts are characteristics, in the sense of what constitutes the knowledge of a thing, but also that they are partial concepts, that is, concepts allowing us the determination of concepts lower than them.

I would conclude this section by remarking that claiming that an object Γ is identical to an object Δ (in the sense of their complete coincidence) means claiming that Γ falls under all the concepts under which the object Δ falls, and the other way around. In other words, Γ and Δ have the same properties. This relation of complete identity is not conceivable in the case of concepts, and in order to avoid this difficulty Frege introduces the notion of 'second level relation' (Frege, 1892–95, pp. 120–4). One cannot speak about concepts in the same way as one speaks about objects. Actually, the 'second level relation' has nothing to do with concepts, but with their graphs, that is, with their extensions (which is the same): 'We could also have termed [graph] the extension of the concept: [for example] square root of 1 [if the function is $x^2 = 1$]' (Frege, 1891, p. 32).

Thus *two concepts are in a second level relation*, that is, in a relation similar to identity between objects, only if their extensions coincide: 'what two concept-words mean is the same if and only if the extensions of the corresponding concepts coincide' (Frege, 1892–95, p. 122).[25]

Frege's concept of 'concept'

Frege's analysis of propositions in terms of both the distinction between argument and function, and concepts as type-II-functions, is certainly more effective than the classical analysis, since it avoids many difficulties. Nevertheless, it has some serious problems.

We can begin the *cahiers de doléance* by mentioning that the theory of the graphs was conclusively refuted by Russell's well-known objection. A second problem is connected with the so-called 'horse paradox'. In a passage, Frege claims that 'the singular definite article always indicates an object' (Frege, 1892b, p. 45). This way of characterizing singular terms is unproblematic in connection with expressions such as 'the ball in my hand' or 'Fido the dog' and so on, but becomes more complicated as soon as one meets expressions such as 'the concept *horse*', 'the concept *Venus*', or, in general terms, 'the concept *F*'. Since these are expressions introduced by the definite article, they should be considered as singular terms referring to objects. Furthermore, they should also be singular terms because they lack the main features of predicates, that is, incompleteness. Nevertheless, what kind of objects should they refer to? Frege confirms that we are dealing with objects of a special kind, those objects called 'concept-correlates' by Wells (1951, p. 17).

However, while they have a name, thanks to Wells, there is still the problem of understanding what kind of objects they are. Furthermore, where are they? In the external world? In the subject's mind? In the third realm of objectivity? Moreover, Frege himself seems to doubt their existence: 'the *Bedeutung* of the expression "the concept *equilateral rectangle*" (if there is one in this case) is an object' (Frege, 1892–95, pp. 119–20).

The problems, though, are not only concerned with Fregean ontology. If we saturate functional expressions such as '*x* is a concept' with expressions such as 'the concept *F*', we obtain false propositions such as,

the concept F is a concept,

since 'the concept F' does not denote a concept, but an object. Moreover, if we saturate with the same expression the unsaturated expression '*x* is not a concept', we obtain a proposition which is self-contradictory: 'The concept F is not a concept'. Frege was well aware of these obstacles: 'it must indeed be recognized that here we are confronted by an awkwardness of language, which I admit cannot be avoided, if we say that the concept horse is not a concept' (Frege, 1892b, p. 46).

Even if it is an unavoidable 'awkwardness', Frege does not consider it as a serious problem that can endanger his proposal as a whole. Actually, perhaps a solution can be found; perhaps under the condition that we do not start from the language, but from the thought, that is, under the condition that we accept the interpretation which I am supporting. From this point of view, the 'awkwardness' is only in the natural language which disguises the thought, but in neither the thought nor in the perfect language isomorphous to the thought. As in the case of the active and passive forms of a proposition, Frege makes clear what the underlying concept is, and thus how the proposition of the perfect language that makes it perspicuous should be. In the case of the paradox in question, perhaps we could assume a similar attitude.

This also allows us to dismiss as wrong Black's comment that since the locution 'the function *f*' starts with a determinate article and thus does not denote a concept, but an object, the language used by Frege to write his papers on function and concept must be completely inconsistent (Black, 1954, p. 242). From this point of view, 'the function *f* is a concept' would be false, as would be the proposition 'the function *f* is a function'. It would then be difficult to understand the *Sinn* of propositions such as 'the function *f* is incomplete': what thoughts do they express? How can objects referred to by 'the function *f*' be incomplete? Are some objects complete and others incomplete? Black might have a good point if we were to consider only natural language, but we know this to be ambiguous and disguising. Frege could, of course, have translated his papers, which were written in natural language,[26] into the perfect language. But this would have undermined the good written style that he obtained by using natural language, notwithstanding its traps. What is important is to understand each other and yet to be aware of the traps.

However, concerning Frege's remark that every singular term has to begin with the definite article, Dummett (1973) points out how extremely unsatisfactory such a pseudo-definition of singular terms is, probably even for Frege himself. It can be applied neither to languages without singular-plural distinction, nor to those without articles. Furthermore, there are some locutions that begin with a singular determinate article which Frege himself – according to Dummett – would not count as singular terms; and there are locutions which do not begin with a singular determinate article, but which would be counted among singular terms by Frege. However, it must be said that Dummett's point would be stronger if Frege had started from a linguistic analysis. In that case, the fact that other languages lack elements that belong to German, or the fact that German can lack elements belonging to other languages, would cause problems.

But there is another point in the Fregean doctrine of concept to be considered; that which allows us to deal with concepts under which only one object falls. The Fregean system explicitly admits concepts with a unitary extension, even if this requires a rather complex procedure. Once more we need to translate a proposition written in the ambiguous natural language into a proposition expressed in the logically clear and enlightening perfect language. In particular, a given singular term 'C' can be transformed into a functional expression such as 'x is no other than C'; the latter is an unsaturated expression denoting a concept under which only one object falls, that is, C. Thus, when we write 'the morning star is no other than Venus', 'what is predicated here is thus not *Venus* but *no other than Venus*. These words stand for a concept; admittedly only one object falls under this' (Frege, 1892b, p. 44). Actually at this point a new problem would arise, even if, to be honest, it would not concern Frege's system, and I shall deal with it later: in what sense can we speak of universality in the case of a concept having a unitary extension?

Further, Frege's proposal of the concept of 'concept' fits into a particular epistemological framework, creating a sense of unease in anyone interested in natural sciences, either as a philosopher, or historian, or scientist. The Fregean perfect language has a very embarrassing characteristic. It can contain neither non-referring singular terms, nor non-referring propositions (cf. Ibid., p. 58 and pp. 62–3; also Frege, 1892–95, pp. 124–5). Both of these are common in the languages of the natural sciences, and, consequently, in the languages of the history and the philosophy of natural sciences.[27]

Nevertheless Frege himself noted that we can grasp the *Sinn* of the singular term 'moly', even if we know that there is no object having the features of the herb which Hermes gave Ulysses to protect him from Circe's magic. Similarly, there can be propositions having a *Sinn* (which we can grasp), but without *Bedeutung*. If a predicate is saturated by a term without *Bedeutung*, it gives a proposition without *Bedeutung* in its own turn. 'Moly is a herb with a black root and a white flower' is a proposition that has a thought as its *Sinn*, but which has no *Bedeutung*. Therefore it is neither true nor false.

In the natural languages, in the languages of natural sciences, and in those dealing with natural sciences, linguistic expressions of this kind are perfectly acceptable, but they are not acceptable in Frege's perfect language, which is isomorphic to the (true) thought. Frege claims that, in the case of poetry, *Sinn* alone can do, but science (which science is he referring to?) requires also the *Bedeutung*, because it is only by means of the *Bedeutungen* that we can speak of truth and falsehood of propositions (Ibid., p. 124).[28]

> The thought loses value for us as soon as we recognize that the *Bedeutung* of one of its parts is missing. We are therefore justified in not being satisfied with the *Sinn* of a proposition, and in inquiring as to its *Bedeutung*. But now why do we want every proper name to have not only *Sinn*, but also *Bedeutung*? Why is the thought not enough for us? Because, and to the extent that we are concerned with its truth-value [...] It is the striving for truth that drives us always to advance from the *Sinn* to the thing meant.
>
> (Frege, 1892a, p. 63)

From this passage, we could wrongly conclude that Frege supports a correspondent theory of truth, but that is not the case. Truth is not a relation between the thought and something external. However, this is another story, and I skip it, even if a philosopher of natural sciences cannot easily maintain Frege's theory of truth.

Finally, it is worth noting another problematic aspect of Frege's view: since the perfect language has to contain only true and false propositions, all the predicates referring to open (or vague) concepts must be eliminated. In other words, one must always be capable of deciding if an object falls under some concept (and, thus, if the corresponding proposition is true), or if the object does not fall under some concept (and, thus, if the corresponding proposition is false). Therefore an open or vague concept cannot be accepted (compare Frege, 1891, p. 33; 1892a,

p. 70; 1892–95, p. 124): 'Truth does not admit of more or less' (Frege, 1918–19, p. 3).

A concept always has to be closed; otherwise it has to be eliminated without any hesitation. Unfortunately, scientists need open or vague concepts, and so do historians and philosophers of science (and philosophers *tout court*).[29] This is not at all satisfying for anyone dealing with sciences of nature, either as a scientist, or as a historian, or as a philosopher of science.

1.1.5 Cassirer, Frege... and Kant

As noted before, Cassirer tried to fuse Kantian transcendentalism together with Fregean logicism. Is such a fusion really plausible? Can we really fuse together Kant's function-concept and Frege's function-concept? What value can such a philosophical manoeuvre have?

Let us start with the question of whether such a fusion is possible, and recall two passages. In the first, already quoted (Cassirer, 1923–29, Vol. 3, pp. 295–6), Cassirer accepts the Fregean notion of *propositional function* mediated by Russell. In the second, Cassirer argues that we have to abandon the classical notion of concept and accept the Kantian one: concept is the unity of rules by means of which a multiplicity of contents is grouped together, that is, concept is *a function of synthesis* (Ibid., Ch. I, § II).

In the former, Cassirer is a reader of Russell, and indirectly of Frege; in the latter, Cassirer is a pupil of Kant. But these are two different Cassirers, who do not succeed in becoming one. First of all, did Cassirer know Frege's works? In Cassirer's works where the question of concepts is mainly stressed (*Substance and Function* and *The Philosophy of Symbolic Forms*), Frege is actually quoted, but only with reference to his *The Foundations of Arithmetic* (1884) and 'A Critical Elucidation of some Points in Schröder's *Vorlesungen über die Algebra der Logik*' (1895). Cassirer does not seem to know Frege's fundamental works of 1891 and 1892. Moreover, from *The Philosophy of Symbolic Forms*, it is not even clear if Cassirer quotes directly from Frege (1895), or if he uses W. Burkamp's work on concept (Burkamp, 1927), which he quotes frequently. Thus, even if some of Cassirer's titles (for example, *Substance and Function*,[30] 'Concept and object') seem to refer to Frege's titles of 1891–92, Cassirer does not appear to be acquainted with them directly. Moreover, he adopts Russell's use of Frege's theory. That is, he considers only that part in which the concept, intensionally interpreted, is seen as an unsaturated function, that, if correctly saturated, gives a true proposition.

From Cassirer's writings, it is not clear if he was aware that even if concepts are functions, not all functions are concepts, since not all functions give truth-values when they are saturated. Concepts are functions only if these are interpreted as type-II-functions. Furthermore, even if in Cassirer's work there is a theory of signs (or rather of symbols), there is no trace of Frege's distinction between sign, *Sinn* and *Bedeutung*. Consequently, the concept is not considered as the *Bedeutung* of an unsaturated expression.

However, more important, especially in relation to what Cassirer is aiming at in his *The Philosophy of Symbolic Forms*, is the question of value: would a strictly Fregean doctrine of concept really be useful for his theory? In his 1923–29 work, Cassirer was trying to extend Kant's approach to mathematics and physics to all fields of knowledge. He was also trying to give a functional account of myth, art, literature and so on, that is, of every symbolic form. Unfortunately, all these fields contain expressions (of both singular terms and propositions) having no *Bedeutung*. Therefore, either one reinterprets in a quite new (and non-Fregean) way the idea both of concept as mathematical function, and of what its saturation is, or an approach of this kind is not only irrelevant, but also inconsistent. On the one hand, one should accept the thesis that a proposition is true if and only if it is obtained through the saturation with arguments having the right *Bedeutung*; on the other hand, one must account for, for example, mythological propositions, which cannot be the result of a saturation with terms having any reference at all. Consequently, had we to accept a strictly Fregean approach, most of the content of *The Philosophy of Symbolic Forms* should be rejected, or considered as poetry. Cassirer, I think would not be happy about this.

This is the main reason behind my claim that in Cassirer's work the notion of 'concept' as mathematical function (in the sense of the type-II-function) is never used, apart from in those pages where the impossible fusion with the notion of concept as transcendental function is attempted. It seems that the juxtaposition cannot be usefully employed to further the aims of the work as a whole. This failed juxtaposition could be a result of Cassirer's critique of the class-concept. As shown, Cassirer starts from a reflection on the foundations of mathematics and, probably, it is this that draws him to introduce the notion of mathematical function into his works. Moreover, the two notions of function (and, thus, of concept) are totally different: *according to Kant, a function is the product of an act of synthesis; according to Frege, it is an operator which maps objects into objects*. They are two completely heterogeneous things. It is true that any mathematical expression can be understood as a function in Kant's sense, as Cassirer pointed out. Nevertheless, this means that it is no longer interpreted as something mapping something into something else, but transcendentally, as a representation of representations. However, a representation of representations cannot be interpreted as a function in Frege's sense, since it is not at all something which maps, but something which unifies.

In conclusion, Cassirer's attempt to fuse together the concept as function à la Kant and the concept as function à la Frege seems to be a failure or, at least, a spurious aspect of his work. What would be sufficient for Cassirer's proposal is the Kantian approach, especially if it were freed – and Cassirer freed it – from the constraints of the original apodicticity. This is a conclusion based on Cassirer's level of analysis, which is transcendental rather than logical, and on his aim, which is not the construction of a perfect language isomorphic to a third quasi-Platonic realm but an account of all kinds of symbolic forms. Furthermore, it is only by the acceptance of the Kantian way, rather than Frege's, that he can reach a pure epistemological (and not logical) conclusion according to which a concept is a hypothetical synthesis, a constitutive representation, and a projection of cognitive significance.

It is along this path that I will move in the next section, where a different theory of concepts will be proposed. That is, I will begin from a Kantian position according to which concept is a function of synthesis, a unity of rules. From this basis, I will argue for a theory of concepts that utilizes a notion – that of 'intentional reference' – which is extraneous to the Kantian tradition. However, it will be this notion that will permit me to reach my (Kantian) aim.

1.2 Concepts as representations and as rules

I will start from scientific language, since it has a vital role in the formation and communication of theoretical and experimental scientific results. It is by means of language that scientific views – which are constructed by appropriately putting together concepts in judgements – can be made intersubjective, can be grasped by anybody, and, thus, judged. Language can be thought of as a third element that objectifies the relations between human beings and the world. It expresses the ways in which human beings constitute the world as an object of knowledge, and the world reacts to humanity by checking and correcting its conceptual products.

However, I will not venture into the debate concerning the relations between thought and language. Neither will I discuss the problem of which came first, nor the argument that one can exist independently of the other. Rather, I wish to begin my proposal from something that, I believe, should not be disregarded: in non-pathological cases, *there is a homomorphism between language and thought, in the sense that everything that can be said, can be thought.* Not an isomorphism, since we should not disallow the possibility that there are products – such as the subjective images of objects – which, even if they can be thought, cannot be linguistically represented, otherwise they would be objectivized and, thus, vanish.

By taking this homomorphism between thought and scientific language for granted, I will move from the latter to grasp some products of the former. However, it is worth noting at the outset that the homomorphism between thought and language cannot be extended to the world. For the world contains both more and less than what is contained in thought. This is not the contradiction it might seem. On the one hand, certainly, the world contains less, since many conceptual products (for example, those of myth, literature, and even of science) have no counterpart in the world, or it can be found that they do not have one. On the other hand, the world contains more, since had each object, or each relation between objects, a corresponding intellectual product, knowledge could not have grown, or could not grow, and this is not the case. To accept that there is a homomorphism between thought and language, or rather *between linguistic representations and corresponding conceptual representations*, is sufficient. Because of this assumption, the first question to be answered is: what are the minimal elements that a linguistic representation must have?

If we analyse any linguistic representation, we recognize that it contains expressions which have to do with individual objects – 'Moon', 'Jove', 'aether' and so on – or collections of objects – 'planet', 'electron', 'adron' and so on. But this is not enough. Linguistic representations also contain expressions that allow us to speak in terms of properties that seem to belong to objects: thus, the electron is said 'to have a charge', the glass 'to be fragile', the body 'to be massive' and so on. Furthermore, there are expressions which allow us to deal with the relations between objects: A 'is heavier' than B, A 'is a less efficient conductor' than B and so on.

Let us pause to consider if there is a difference between property and relation. Is it true that properties are 'intrinsic' (in whatever sense you want) to the objects possessing them, whereas relations are 'extrinsic' (in whatever sense you want)? Let us consider the following passage from Helmholtz:

> Actually each property or quality of a thing is nothing but its capacity to act to produce certain effects on other things [...] We call such an activity a property when we hold the reagent, on which it works, in mind as self-evident without naming it. Therefore we speak of the solubility of a substance, that is its reaction to water; we speak of its weight, that is, attraction for the earth; and with the same justification we call it blue, since we thereby presuppose as self-evident, that we are merely concerned with defining its effect on a normal eye. However, if what we call a property always involves relation between two things, then such an action naturally never depends on the nature of one of the agents alone, but always exists only in relation to and in dependence upon the nature of a second object, on which the effect is exerted.
>
> (Helmholtz, 1868, p. 321)

From this passage, we realize that properties can also be considered as relations, even if relations of a certain kind. The object O, which is said to possess the property P, can be thought of as an object in a given relation with the catalyst object O′, that is, with an object that, in certain well-determined empirical situations, permits us to detect the property P possessed by O. From this point of view, the property P is not 'intrinsic' to O, but something which emerges from a particular empirical relation O and O′. While the presence of O′ is clearly emphasized at the empirical level (P could not be detected without O′), it can be neglected at the theoretical level. It is here that we can find the source of the erroneous idea that P is something possessed by O independently of anything else, that is, something completely un-relational. The electron is said 'to be charged', that is, it possesses the property P = 'being charged', but this does not mean that this is a non-relational property. Actually, being charged is something detectable only when the electron enters a relation with another object of a particular type, that is, the catalyst object, for example a proton, an ion, a current. At the empirical level, without such a proton, ion or current, there is no reason to claim that the electron is charged. Nevertheless, we can neglect the proton, ion or current at the theoretical level.

Therefore, instead of speaking of properties and relations as two different things, it would be better to speak of *implicit properties* and *explicit properties*. The *implicit property* of an object O is what emerges from a well-determined empirical relation between the object O and a well-determined catalyst object O′, whose presence is necessary at this level even if it can be negligible at the theoretical level.

The *explicit property* of an object O is what emerges from a well-determined empirical relation between the object O and a well-determined catalyst object O′, whose presence is necessary both at this level and at the theoretical level. However, from now on, I will simply use 'property', both for the implicit properties and for the explicit properties which, as we have seen, are both relational.

Returning to our point, we have expressions related to individual objects, expressions related to collective objects, and expressions related to properties. Of course, these three classes are not sufficient, since a linguistic representation is not a mere juxtaposition of their elements. It is rather something in which objects and their properties are connected in propositions. It is possible to speak about objects and their properties in at least four different ways:

(1) we can speak about the properties possessed by collections of objects: 'All protons are adrons', 'All quarks are confined', 'All moving magnets cause a current in the close conductors';
(2) we can speak about the properties possessed by individual objects: 'The Moon is not a radioactive source', 'Mars has a mass which is about 0.11 times the Earth's';
(3) we can speak about the properties possessed by singular objects belonging to classes of collective objects: 'This body has a mass of 8 kg', 'This conductor has a resistance of 5 Ω'; 'The magnet on the blue table generates a field stronger than the one generated by the magnet on the red table';
(4) we can speak about the existence of some individual object and some singular instances of collective objects: 'There is a planet which has a diameter of 6780 km', 'There is a magnetic single pole', 'There exists a 3 m long body', 'There exist a 4 Mev particle accelerator'.

At this point we could be satisfied, since we have all we need in order to construct a (minimal) scientific linguistic representation. That is, we have expressions about individual and collective objects, and non-quantified or quantified (in a universal or existential way) expressions speaking of objects and properties.

These classes belong to the linguistic side of a homomorphism, whose other side is made up of conceptual products, in particular representations. As a consequence, there is a correlate conceptual product (that is, a concept and a judgement) for each class of linguistic expressions. That is, there are (1) concepts corresponding to the expressions relative to individual and collective objects and (2) non-quantified and quantified judgements corresponding to non-quantified and quantified propositions. However, now let us put aside judgements, and consider concepts.

The presence of concepts of individual objects could come as a surprise, although, after a short and intuitive reflection, this should not seem so bizarre. Frege showed there can be concepts under which only one object falls. In the perfect language, which does not disguise the thought, there are unsaturated expressions like 'x is no other than Venus', whose reference (à la Frege) is precisely a concept having an extension with only one object, that is, the planet Venus. Moreover nobody questions the existence of concepts with infinite extensions – 'man', 'dog', 'potato' and so on – or the possibility of concepts with finite extensions either – 'the planets of the solar system', 'the flowers of John's garden', 'the stars of the Milky Way' and so on. In this case, the cardinality is given by *n*, finite number. Yet, I hope, nobody would be keen to suggest that *n* must necessarily be greater than 1. Do we really want to eliminate $n = 1$ or $n = 0$? For what reason? Nevertheless, a problem arises: in what sense can we speak about universality in the case of concepts having unitary or empty extension? This question will remain unanswered for a while: before dealing with it I will make some other steps.

However, now is the right time to express and argue my thesis. I will proceed by presenting and discussing a chain of partial theses. This means that only at the very end should the theoretical frame become clear.

1.2.1 The proposal

Thesis I: Concepts are representations and rules

I propose to think of the concept as a *function of synthesis* due to a unifying intellectual act allowing us to group together several representations under a unique representation. We do not infer from objects what is common; otherwise we should face the problems of the class-concept. Rather, since the intellectual act produces the function of synthesis (that is, the concept), it produces what unifies several objects, that is, what will be common among them thereafter. In this way, the concept represents the objects that fall under it, that is, the objects it unifies. Here it is the notion of *concept as representation*.

Nevertheless, the synthetic intellectual act does not only produce the representation, but, at the same time, produces the rule that must be satisfied by an object so that it can be considered as an object falling under that representation. Therefore, at the very moment in which we have the representation, we also have the rule indicating what has to be represented. Here it is the notion of *concept as rule.*

The two aspects are totally inseparable. There cannot be representation of something without a rule that indicates what can be represented. Vice versa, there cannot be a rule of synthesis without a representation representing what is synthesized.

Therefore, *the concept can be considered as the representation of what is common to what it synthesizes, and as the rule which identifies what is common.*

Let us take the concept of 'gene'. As a representation the concept of 'gene' synthetically represents all the objects satisfying the rule intrinsic to that representation: 'sequence of nucleotides belonging to a genome that works as genetic unit, and can be transcripted into a RNA sequence'; namely, all the objects satisfying the concept of 'gene' as a rule.

Speaking about *concepts as representations* and *concepts as rules* means speaking about the two epistemological sides (facing two epistemological directions) of the same coin. In particular, the concept as representation deals with the reference of the concept, while the concept as a rule deals with the sense of the concept.

Thesis II.1: The concept as rule is connected to the sense of the concept as such

What does possessing a rule mean? Nothing but knowing how to apply it and, thus, knowing what satisfies it; that is, what can be represented by the representation which intrinsically contains it in order to be that representation. Thus, possessing a rule means possessing a concept. In other words, grasping the rule means grasping the concept, more precisely its sense. I know what the sense of the concept 'gene' is, if I know the rule permitting me to subsume certain objects under that concept. Note, however, that a concept is not really only one rule, but any concept must be thought of as a unity of many rules: all those rules permitting the knowing subject to put the right objects under that concept. Let us think of the concept of a collective object such as 'electron'. Actually, we should not speak of the sense of the concept 'electron' as a rule, but as a *unity of rules*. Grasping the sense of the concept 'electron', in fact, requires grasping a set of rules synthesized by that concept: 'an electron is the source of an electrostatic field', 'an electron is an elementary charge', 'an electron has a charge of 1.6×10^{-19} C', 'an electron has a mass of 9.1×10^{-31} kg' and so on.

In this case, grasping the rules allowing us to grasp the sense of the concept 'electron' means grasping classical electromagnetism. But can other rules be given? Of course. I could have given some other rules, like 'an electron has spin of $\frac{1}{2}$', 'an electron has a mass of 0.5 MeV', 'an electron is a stable particle', 'the anti-particle of the electron is the positron' and so on. In this way, we would still have a set of rules synthesized by the concept 'electron', but such a concept would no longer have the same sense as the previous concept, since the rules would be given not by classical electromagnetism but by quantum mechanics.

Thus, the concept 'electron' as a rule is really the unity of the rules synthesized by it, and to grasp that unity is to grasp its sense.

Definition: By *semanticizing area* I mean the set of rules, synthesized in a concept, which are sufficient to grasp its sense

Changing the semanticizing area (for example, from classical electromagnetism to quantum mechanics) means changing the sense of a concept. Thus, the sense of a concept is not something belonging to that concept as such, but something belonging to the whole set of rules synthesized in that concept.

Possessing several different semanticizing areas means grasping several different senses of the concept. Not only that: the wider a possessed semanticizing area is, the wider the sense of the possessed concept is.

Let us suppose we do not have any semanticizing area allowing us to grasp the sense of the concept 'daturine'. We could remain in that cognitive state, that is, we could never know what 'daturine' is. Let us suppose, instead, we know that daturine is contained in some plants, the daturas, and that it is an alkaloid. Now we have an extremely narrow semanticizing area, since we know, more or less, what 'to be contained in plants' and 'alkaloid' mean. In this case, we can already claim that we have grasped the concept of 'daturine', even if in an extremely poor and shallow way. In order to grasp the concept in a richer and deeper way, we should acquire, for example, some rules of the chemistry of alkaloids. Similarly, if we can associate the concept 'ammonia' to a semanticizing area containing rules such as 'ammonia is soluble in water', 'ammonia is used in refrigerators', we can already say that we have grasped something of its sense. However, we can enlarge that semanticizing area (and thus enlarge the sense of the concept) through possession of rules like 'the chemical formula of ammonia is NH_3', 'ammonia is a gas', 'ammonia liquefies at 10°C of temperature and at 6 Atm of pressure', 'ammonia combines with acids and originates ammoniac salts'. Therefore, we have the following corollary.

Corollary: Possessing different semanticizing areas means grasping different senses of the concepts; possessing wider and wider semanticizing areas means grasping more and more deeply the sense of the concept

These lead us to the conclusion that the semanticizing area is not unique, and that it cannot be something clearly delimited or delimitable, especially if we are dealing with everyday concepts. The limits of a semanticizing area are, in many cases, subjective limits, in the sense that each individual has his own particular extensions of a given semanticizing area.

Note that if this is valid, the aim of teaching should be threefold: (1) to standardize the students' semanticizing areas (to create a common base of knowledge); (2) to propose new ones (to allow them to know more of the world); (3) to enlarge those they already possess (to give them the opportunity to know the world more deeply).

Thesis II.2: A concept as representation is connected to its reference

Whereas the sense of a concept concerns the rules the latter unifies, the reference of a concept regards its being a representation of something different from itself, that is, a representation of its reference.

Thesis III: Every concept refers intentionally to an object

Each representation *qua* representation is representation of something different from itself, in particular it is a representation of its reference, even if it is not given that each reference is something existent.

At this point, a distinction must be made between *intentional object* and *empirical object*.

Definitions

Given a concept of an object (an object-concept):

(1) By *intentional object* I mean its intentional reference, that is, what it refers to conceptually. In this way the intentional object has all the determinations distinguishing it, that is, it has an essence – to be intended only in the epistemological sense, and therefore without any metaphysical commitment. Note that the intentional object has nothing to do with the subjective image of an object. As I will show, it is completely objective.
(2) By *empirical object* I mean its empirical reference, that is, what it refers to in the world. In this way, the empirical object has all

the determinations distinguishing it, and it is also thought of as existing.

Thus, existence is not something concerning all objects, that is, intentional objects, but only concerns empirical objects. It is important to avoid the fallacy of prejudice about the existing.

Warning: The fallacy of prejudice about the existing
The prejudice about the existing arises from the fact that often we do not realize that what exists, or existed, or will exist, has only a (non-empty) intersection with the set of what we can know. There are existing objects we do not know yet, but there are also objects we know, even if they do not exist. We do not only know bar counters, bendable seats, slicing machines, grapes, electrons, molecules, which exist, existed, and probably will exist. We also know luminiferous aether, phlogiston, the Seven Dwarfs, Ulysses, Nausicaa, which never existed in the past, do not exist now, and very probably will not exist in the future. Reducing all knowledge to what was, is, or will be empirically known or knowable, amounts to unwarrantedly ignoring numerous products of human thought, which could not, cannot, or will ever be empirically known, or knowable.

I will skip the (neither trivial nor marginal) problem concerning what empirically knowable means, since this is not the problem in question. Instead, it is worth asking: should we really constrain our knowledge to what is empirically knowable? Should we eliminate, or stop discussing, fictional objects, or mythical objects? We do not really know Snow White, Ulrich (the man without qualities), the hippogriff, Chiron the centaur or Apollo? Should we really eliminate from our speech physically, or logically impossible objects? We do not really know the golden mountain or the fused ice, the circular square or the convex concave?

Here, what 'to know' means must be clarified. First of all, *'to know' must not be interpreted as synonymous with 'to know empirically'*. This would mean supporting a *positivistic parody of the worst positivism*. In particular, to constrain the realm of knowable to what is empirically knowable would jeopardize the possibility not only of speaking cognitively about what will never exist, but also about what might found be to exist in the future. However, should the use of 'to know' in the case of intentional objects prove disturbing, we could distinguish between *intentional knowledge* (regarding intentional objects) and *empirical knowledge* (regarding empirical objects).

Having marked the prejudice about the existing, I can go on to the following principle.

Principle of independence of essence from existence: That any intentional object has individual and well-defined determinations, that is, an essence, that is completely independent of its existence
Basically, this principle summarizes something we have already claimed: that we can know essences of objects, without concerning ourselves with their possible existence. In other words, intentional knowledge is completely independent of empirical knowledge.

Thesis IV: Not every object-concept also refers to an empirical object
Since it is a conceptual representation, every object-concept intentionally represents an object, but this is not necessarily also an empirical object. If we want to avoid the trap of the prejudice about the existing, we should always make clear if we are intending either what is indicated and really exists (the empirical object), or what is simply indicated (the intentional object). In the first case, we have to do with an *empirical reference*, and there will be something existing represented by it. In the second case, we have to do with an *intentional reference*, and what is represented does not have existence. Thus, the following.

Corollary: Every representation has an intentional reference, but not all representations have an empirical reference
Let us start with concepts of collective objects, specifically the concept of 'planet'. This certainly has an intentional reference, that is, the intentional object 'planet' which can be known intentionally. However, there is no empirical reference, since there is no empirical object 'planet' that can be empirically detected. Here a misunderstanding could arise: *there is a particular planet*, in the sense that a particular planet exists, *but the planet qua planet does not exist.* Thus, every collective object-concept has an intentional reference, which is the intentional object it represents, but it does not have an empirical reference, since empirical references are always individuals. Moreover, a singular object-concept, obtained by particularizing a collective object-concept, can have an empirical reference. Of course, it still has an intentional reference. For example, the concept 'this electron' (obtained by making the collective object-concept 'electron' particular) has an intentional reference (which is the intentional electron in question), but also an empirical reference: the empirical electron we are referring to.

The same analysis can be made in the case of individual object-concepts. Whereas, the individual object-concept 'Chiron the centaur' has only an intentional reference, since there is no empirical object corresponding to Chiron the centaur, the individual object-concept 'Venus the planet' has both an intentional reference and an empirical reference (the empirical object corresponding to Venus the planet).

Thesis V: Property-concepts are epistemologically prior

When I proposed an example of the way in which the sense of a concept is given, by means of the rules of its semanticizing area (rules which are synthesized by the concept itself, since it is the unity of rules), I claimed that the sense of the concept 'classical electron' is individuated by the rules 'an electron is the source of an electrostatic field', 'an electron is an elementary charge', 'the charge of an electron is 1.6×10^{-19} C', 'an electron has a mass of 9.1×10^{-31} kg', and so on. By examining these propositions, we can conclude that to grasp the sense of the concept 'classical electron' we need to master the senses of property-concepts such as 'being an elementary charge', 'being the source of an electrostatic field', 'having a charge of 1.6×10^{-19} C', 'having a mass of 9.1×10^{-31} kg', and so on. Thus, grasping the sense of an object-concept amounts to grasping the correlated semanticizing area, that is, the properties pertaining to the latter. Therefore, since properties are attributed in propositions, it means also grasping the propositions belonging to the same semanticizing area.

Thus, grasping the sense of the object-concept can involve different things, according to what we want to emphasize. It can mean:

(1) grasping the semanticizing area in which that concept is contained (emphasis on the holistic approach to the sense);
(2) grasping the rules belonging to that semanticizing area (emphasis on the concept as rule);
(3) grasping the properties possessed by that intentional object, in accordance with the rules belonging to that semanticizing area (emphasis on properties).

As noted, grasping the sense of a concept also involves the individuation of its reference. Similarly, grasping the property-concepts, which are the characteristics of that particular object-concept in that particular semanticizing area, involves the individuation of the references of the object-concepts by means of the references of property-concepts (that is, the properties possessed by the denoted intentional object).

Furthermore, since there are intentional references and empirical references, there will also be *intentional properties* (possessed by intentional objects) and *empirical properties* (possessed by empirical objects).

From an epistemological point of view, the distinction between empirical and intentional properties is very important. The intentional properties possessed by intentional objects are determined by the semanticizing area to which those object-concepts and those property-concepts belong, while the empirical properties are empirically determined by means of experiments. Therefore, what allows us to know which aspects of an object are empirical is resort to the world. This does not amount to an assertion that the empirical properties are inferred from experience; rather, experience allows us to determine which intentional properties also have empirical counterparts. Thus:

(1) grasping the property-concepts of a semanticizing area means grasping the object-concepts belonging to that area;
(2) knowing the intentional references of the property-concepts means knowing the references of object-concepts possessing those intentional properties;
(3) knowing the empirical references of the property-concepts means knowing the empirical references of the object-concepts possessing those empirical properties.

Therefore, *I intentionally know an intentional object because I know the intentional properties it possesses, and I empirically know an empirical object because I know the empirical properties which it possesses*. In other words, the intentional object is given by the intentional properties it possesses: it is a *knot of those properties*, and its corresponding object-concept synthesizes (as the unity of rules) exactly those properties. Analogously, the empirical object is given by the empirical properties it possesses: it is a *knot of those properties*, and its corresponding object-concept synthesizes (as the unity of rules) exactly those properties. Hence the following.

Corollary: An intentional object is a knot of intentional properties; an empirical object is a knot of empirically properties
This is the reason why I claim the epistemological priority of property-concepts, and thus the epistemological priority of their references (that is, properties), especially of the intentional ones.

Thesis VI: Only singular empirical properties of objects can be empirically detected, but not empirical objects as a whole
When I mentioned the empirical properties, I affirmed that they are what allows us to speak about the existence of the object possessing them. But, of course, *we can empirically detect only individual properties of the objects, not objects as a whole*. It would be a mistake to claim we have detected an object as a whole. There is no experimental apparatus allowing us to detect objects in their completeness. We can detect that an object is lucid, transparent, acid, radioactive, heavy, charged, with spin $\frac{1}{2}$ and so on. *The empirical object that we empirically know is the knot of the empirical properties that we can empirically detect, one independently of the other*. Furthermore, and we should keep this in mind, it is not necessarily true that all intentional properties have an empirical counterpart. This is so only of those that can be identified by means of an experimental apparatus. This means that the existence we can discuss can only be given by the properties we can directly, or inferentially, detect. Once more, these are the only ones that we know empirically. Note that we can have an intentional object made up of n intentional properties $(p_1^i, p_2^i, \ldots, p_n^i)$, but whose empirical counterpart is made up of $m \leq n$ empirical properties $(p_1^e, p_2^e, \ldots, p_m^e)$. This can offer a satisfying solution to many problems, as can be seen by answering some of the objections it could meet.

Objection I: What does knowing an intentional object mean, or what does knowing intentionally mean?
I repeat: knowing an intentional object means knowing its intentional properties, and it is precisely such knowledge that makes knowledge of the empirical object possible. We can empirically know empirical properties (that is, we can scientifically know the world around us) because we have conjectured their existence, that is, because we have produced an intentional object possessing intentional properties that cognitively constitute those aspects of the world that we consider as the correlated empirical properties.

Objection II: Is the introduction of intentional objects really necessary? In this way, do we not enlarge ontology unnecessarily?
Certainly, the introduction of intentional objects causes the enlargement of the ontology, but such an enlargement is necessary, for at least two reasons. The first concerns knowledge in general; the second concerns scientific knowledge. Without introducing intentional objects, how could we account for the possibility of knowing of objects that do

not, or cannot, exist? Let us think about Diana, goddess of hunting. She is not an empirical object, nevertheless I know her: Diana is a knot of intentional properties which are the intentional references of property-concepts contained in the semanticizing area provided by classical mythology. Some anthropometrical properties, some amatory properties, some athletic properties, some kinship properties can be attributed to Diana. In the same way, Pegasus, the Seven Dwarfs, Moby Dick and Captain Ahab are intentional objects since they are knots of intentional properties, that is, since they are intentional references of property-concepts belonging to mythological or fictional semanticizing areas. What about W.v.O. Quine's (1948) round-squared cupola of Berkeley College, the concave convexity, the golden mountain, the boiling ice? Do we not know these objects? I claim we do, at least in the sense of intentional knowledge, since we know the intentional properties they possess, and these can easily be inconsistent. In conclusion, we can know them since they are the intentional references of the object-concepts of which the property-concepts – which refer to the properties possessed by them – can be predicated. I can grasp the sense of these odd concepts because I grasp the semanticizing area they belong to. Are these empirical objects? Certainly not, but this is a different problem altogether. Only if we admit the possibility of intentional objects can we account for the (intentional) knowledge of fictional, poetic, mythological, and even physically and logically impossible objects.

Moreover, intentional objects are necessary for scientific knowledge, since it is precisely science which puts possible objects into a sort of limbo. As said before, *first* we produce intentional properties by intellectually producing property-concepts denoting them, and *then* we check if those properties have an empirical counterpart as well. In this way one produces an intentional object, leaving the possibility that there is also an empirical correlate open. Thus, the object is constituted for empirical knowledge. The production of an intentional object by means of its concept (that is, by means of the property-concepts which can be predicated of it) can be thought of as a condition for the possibility of the empirical knowledge of the correlated empirical object, which is, thus, cognitively constituted.

Objection III: What is the essence of an intentional object?
In order to answer this question, we can recall Kant's passage:

> [...] however, we must in no way think of the *real* or *natural essence* of the things, into which we can never gain insight. For since logic

> abstracts from all content of cognition, consequently also from the matter itself, only the *logical* essence of things can be under discussion in this science. And into this we can easily have insight. For to the logical essence belongs nothing but the cognition of all predicates in respect of which an object is determined by *its concept*; the real essence of the things (*esse rei*), on the other hand, requires the cognition of those predicates on which depends, as determining grounds, everything that belongs to its existence. If, for example, we want to determine the logical essence of a body, we do not have to search out the data for this in nature; we only need to direct our reflection to those characteristics which as essential elements ([*nota*] *constitutiva, rationes*) originally constitute its basic concept.
>
> (L, p. 67)

Here, Kant emphasizes the distinction between *logical* (I prefer to call it *epistemological*) *essence* and *real essence*. This distinction should not be forgotten, for too often we tend to consider 'essence' in a strongly metaphysical way. Essence, however, is that something that determines an intentional object *qua* that intentional object, but it has nothing to do with possible existence. *Essence, once more, is what determines that individual intentional object qua that individual intentional object, and not that individual empirical object qua that individual empirical object.* Concerning this, let us keep in mind the principle of the independence of essence from existence.

Objection IV: In what sense can we speak about the universality of a concept?

After warning about the possible misunderstanding of the term 'essence', it is easier to deal with the problem of the universality of concepts, which has been postponed twice: in what sense can a concept having a unitary extension, or a null extension, be considered universal?

To find a satisfying solution, we should go back to the scholastic distinction between *direct universality* (*intentio prima*) and *reflected universality* (*intentio secunda*). By 'direct universality' is meant (and I mean) precisely the essence of the object intentioned by an object-concept. In such a way, that intentional object has to be considered only in its epistemological essence (as a type-object, or, as it is sometimes called, an absolute object, or a conceptual object), that is, independently of its possible particular individual instantiations (its token-objects), and independently of the possibility of also having empirical counterparts. A direct universal is also independent of singularity or multiplicity, and of existence and non-existence. By 'reflected universality' is meant (and I mean) what results from reflecting on the relations between a given essence and the act of determining some individual intentional and empirical objects. In other words, the reflected universality concerns the relation between an essence and the possible individuals that are what they are since they have that essence. Consequently, the reflected universality has an essential content, and it is characterized by the possibility of being in relation of predicability. Therefore, it is only the reflected universal that is apt to be predicated of a multiplicity.

Bearing this distinction in mind, it should not be particularly surprising that I speak about universality also in cases of concepts having unitary, or empty, extensions. I do not refer in this way to the reflected universality, but to the direct universality. All object-concepts are direct universals, since all of them are representations of the correlated intentional objects, that is, all of them have an essence. Nevertheless, not all of them are reflected universals, since not all of them are predicable of a multiplicity. Consequently, the corollary is as follows.

Corollary: From an intentional point of view, no concept is empty
It is misleading to claim that an empty concept is a concept under which no element falls. For all concepts always refer at least to intentional objects. Therefore, from this point of view, no concept is empty, since an intentional object always falls under it. Of course, if we look at the problem from an empirical point of view, things change: a concept can be empty in the sense that no empirical object falls under it. So, one should be clear when speaking of empty concepts. Is one talking from an intentional point of view? In that case, there are no empty concepts. Is one talking from an empirical point of view? Then, there can be empty concepts. Let us note that, (1) from the point of view of direct universality, every concept cannot but be a unitary concept and (2) from the point of view of reflected universality, it could be *n*-ary, where $n = 0, 1, 2, \ldots$

Explication: The difference between characteristic and property
The remarks above open an alternative way to deal with the difference between the notions of 'characteristic' and of 'property', which afflicts the classical theory of concepts in which there is no clear distinction between 'object' and 'concept', and what their features are. However, this ambiguity is latent also in modern (Kant's) and contemporary (Frege's, Cassirer's) theories, even if in those cases the distinction is

clearer. As we have seen, Kant does not pay much attention to the notion of 'property', even if in his *Logic* he considers the notion of 'characteristic', which he takes over from the classical tradition and considers under the two headings of partial concepts and self-standing concepts. In Frege, the distinction between 'property' and 'characteristic' becomes sharper, as a consequence of his sharper distinction between 'concept' and 'object': a property has to do with an object, a characteristic has to do with a concept. However, both a property and a characteristic are concepts; indeed they can be the same concept considered from two different points of view.

In my proposal, properties and characteristics are two different things altogether. This is due to the distinction between the notion of 'property-concept' and that of 'property' (interpreted as reference). In this framework, the property-concepts can be predicated of object-concepts and, thus, *a property-concept is the characteristic of an object-concept, which then is a concept subordinated to it.* Moreover, *a property, since it is a reference, is what is possessed by an object, which is, in turn, a reference of an object-concept.* So, the characteristic belongs to the representation level, the property to the (intentional or empirical) reference level. In the statement 'magnets attract iron' this double aspect is made clear. On the one hand, the property-concept 'attracting iron' is predicated of the object-concept 'magnet', and, thus, the former is a characteristic of the latter, whereas the latter is subordinated to the former. On the other hand, the references of the concept 'magnets' (that is, the objects which are magnets) possess the property referred to by the concept 'attracting iron', that is, the property of attracting iron.

Consequently, a proposition shows one of the rules by means of which grasping the sense of the given object-concept is possible in that semanticizing area, but it also shows that characteristics and properties are at different levels. In conclusion, *an object is a knot of properties and an object-concept is a knot of characteristics* (the property-concepts it can be predicated of). If we eliminate all properties, we will not have any object left, and if we eliminate all the property-concepts, we will not have any object-concept left, exactly in the same way (by using a 'vegetable' metaphor) we do not have any artichoke left, if we remove all its leaves, or we have no onion left, if we remove all its layers (Figure 1.1).

Let us note, though, that reducing an object-concept to the property-concepts predicated of it, or an object to the properties it possesses, does not amount to idealistically evaporating the object. It only amounts to the claim that at an epistemological level an object-concept can only be grasped by grasping the property-concepts of the semanticizing area it

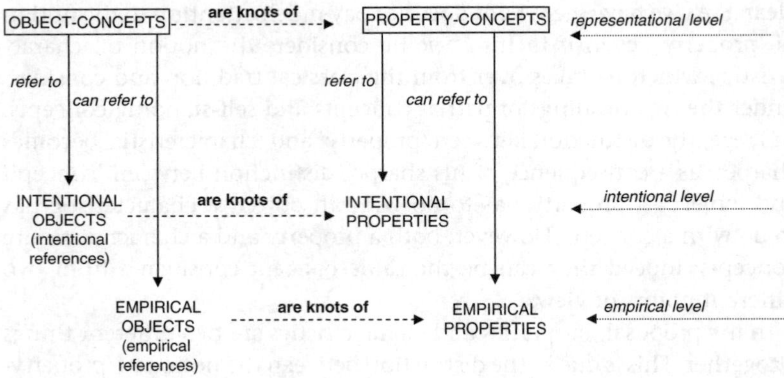

Figure 1.1 Objects, concepts, properties

belongs to. Similarly, an object can only be known by knowing its properties. This has nothing to do with a theory that claims the vanishing of objects at a metaphysical level.

1.3 Reality

Now, let us tackle the question of realism – but in full awareness that *reality, objectivity* and *observability* are three totally different notions, which, unfortunately and often with misleading consequences, have frequently been considered as equivalents. But what does it mean to be a realist? Prima facie, it could seem that it means to believe, or to assume, that an external world exists. Nevertheless, Berkeley (1713, §§ 34-41) confuted this claim. Berkeley, the idealistic philosopher *par excellence,* did not deny reality at all, even if he claimed that *esse est percipi*. Rather, he denied the metaphysical existence of an external world and, therefore, its independence of the knowing subject. It follows that one could be an idealist and a realist at the same time: it is sufficient to suitably define the terms 'idealism' and 'realism'. Already from this example, it is obvious that terms such as 'idealism' and 'realism' have a lot of different senses, and that we should be extremely careful as to how they are used.

However, in this section I propose a realistic approach starting from only one assumption: *there is a primacy of the representations, that is, of the semanticizing areas*. This epistemological claim marks my acceptance of Kant's 'Copernican revolution', according to which we are able to empirically know only what we have a representation of. On the basis of this claim, I will argue for a particular kind of realism that I call *wise realism*, since, as I will discuss, even if I admit that there is something, I suspend judgement on what actually there is, and on how this is actually structured. In other words, I am proposing a kind of *wise realism* based on a metaphysical presupposition concerning the existence of something independent of the knowing subject's mind, but I avow *epoché*, I suspend judgement on how this metaphysical something is made. It seems to me that this position is reasonably plausible, being grounded in the idea that we can claim something interesting only on what we really can be acquainted with, both in the intentional and in the empirical sense. At this point, it is worth recalling that both the metaphysical idealist (maintaining that everything is the knowing subject's construction) and the metaphysical realist (maintaining that there is something independent of the knowing subject's mind) admit that there is something, even if they disagree as to the extension of this something. Both of them admit that mental events exist.[31] According to the metaphysical idealist, these are the only 'things' that there are, whereas according to the metaphysical realist there is much more. However, each kind of metaphysical realist states a different extension of this 'much more'. Some assume that in this 'much more' there are also mathematical entities. Some deny it. Others believe that there are only chairs, tables. Still others include theoretical entities. And so on. The 'something actually existing' is a country populated at times by few entities, at times by many entities: it depends on the particular metaphysical commitment one accepts.

Unlike the metaphysical idealist I assume that there is something independent of the knowing subject's mind. But unlike the many possible ways of being a metaphysical realist, I suspend judgement on how this something is. Therefore I commit neither to the extension nor to the structure of this something. In this way, my epistemology is unavoidably based on metaphysics (there is something), but my *epoché* means I do not require any further precise and explicit commitment.

Note that *the question of the existence of an object is tackled differently by physics, epistemology, ontology and metaphysics*. That is, we could answer it as physicists, as epistemologists, as ontologists and as metaphysicians, that is, from four different points of view allowing us four different answers at four different levels. I wish to answer only as an epistemologist, who is concerned with a theory of how to know, and as an ontologist, who is concerned with a theory of what is known and knowable. Nevertheless, if one accepts the approach here proposed, that is, the Kantian one, many ontological problems are reduced to epistemological problems: the known and knowable are only what we are able to produce representations of. Therefore, I will leave to the physicists the question of the actual empirical detectability of what is known and knowable, and to the metaphysicians the question of their actual status. What sophists and sceptics have challenged since the beginning of epistemological reflection is that we can know the essence of things. Despite such criticism, the idea that representations mirror the object in itself have pervaded the entire history of philosophy and crushed the theory of knowledge into the theory of what is known, or, rather, into its realistic version. The essentialists, both empiricists and rationalists, considered that their aim was to unveil the world in itself and to arrive at a true representation, isomorphous to such a world.

Yet the essentialist empiricists were wrong, because they believed that the original source of knowledge was bare perception. They had the problem of transforming a subjective perception into objective knowledge, which – moreover – had to be regarded as the product of a passive receptivity. Further, at least for some authors, the privilege given to the perceptive moment also involved the problem of justifying generalizations, which had to preserve the truth of what was known to be true only for a few cases. Thus, yet another problem arose: the justification of a chimerical inductive process that they had necessarily to invoke. The criticism of this path – by means of the objection that what we know through the senses is only appearance, and therefore different from what actually is – eliminated the crushing of the theory of knowledge into the theory of what is known. It also created a gap between what really is, but what we cannot know – the noumenon – and what can be known, but what is only appearance – the phenomenon. As a consequence, a differentiation had to be made between representations and the world in itself, and thus the truthful status of representations was weakened, since true knowledge was only knowledge of the world in itself.

Some authors found such a weakening intolerable. Analogously, the abyss between the subjectivity of perception and the objectivity of representation was also intolerable. Such epistemological dualism between the noumenic world and the phenomenal world ought not to entail the loss of true knowledge. Truth had to be recovered, and this process started not from the empirical world, at this point depreciated to pure appearance, but from the other side, that is, from the side of the knowing subject's understanding.

It was Kant who tried to fill the gap between sensibility and understanding and to save the possibility of apodictic knowledge. We know very well the cost of his enterprise. On the one hand, the need for a canonical and binding categorial apparatus whose truth had to warrant the truth of knowledge. On the other hand, the artifices and obscurities of the transcendental deduction which had to eliminate the gap between the passivity and receptivity of sensibility and the activity and objectivity of understanding. However, by the end of the nineteenth century the truth and definiteness of the Kantian categorial apparatus had exploded into thousands of categories considered only as possible and tentative conjectural representations.

Moreover, in order not to transform the application of the categories into a radical form of constructivism, there had to be a connection between the phenomenal world and the noumenic world. According to Kant, this connection existed only insofar as there was an undeniable stimulus by means of which the noumenon activated the passivity of receptivity. Yet, as, for example, Schopenhauer observed in *The World as Will and Representation* – following Jacobi and Schultz – this stimulus was nothing but a causal connection surreptitiously introduced. For how can I state that there is a causal connection between the noumenic world and the phenomenal world? It could be in the noumenic world: but this should be not possible, since I cannot know anything about it. It could be assumed a priori: but we would have the consequence that all of knowledge would be a construction. Therefore, what to do?

This is the very problem of Kantian epistemological dualism: the relation between what can be known, but that cannot warrant the truthfulness of knowledge, and the unknowable, which has to be postulated to avoid falling into radical constructivism. Whoever embraces a position in which there is some kind of epistemological dualism must explain how representations and a world independent of them are connected, and how the validity of the former is justified. This problem has become even more important in the Kantian tradition, in which the certainty allowed by the apodicticity of Kant's categorial set-up has been denied. Consequently, there has been a loss of reliability in the representations. They can no longer be qualified as true representations either by starting from the bottom – since their empirical verification is an impossible dream coming up against the endless number of necessary confirmations – or by starting from the top – since we do not have a set of true principles from which they can be deduced. Consequently, the representations, having lost their centrality, have been considered by some authors as instruments, as conventions, as attempts made to save the phenomena and so on. The explosion of the Kantian categorial apparatus has transformed epistemological dualism into an insane race towards the destruction of the cognitive value of the representations.

There were those who glimpsed a way out of the epistemological dualism in considering the world as negligible and remitting everything to intellectual activity. Here, though, the dilemma is between the horn of nativist rationalism and the horn of constructivist rationalism. According to the first, knowing is unveiling: not unveiling the world in itself, but unveiling the ideas that are in the mind (or in some sort of 'third world'). Moreover, such ideas must be true, and therefore they must be the archetype of what there is in the world. This means that both their existence in the mind (or in some sort of 'third world'), and the existence of a method to reach them must be justified. But, as far as I know, nobody has been able to argue ineluctably either of these. The horn of constructivist rationalism I see as the paradoxical and desperate outcome of the criticisms against the truthfulness of the senses. However, I find it is very difficult to accept the dissolution of the materiality of the world into what the representations are supposed to construct. I find it is very difficult to accept that everything around us is intellectual *invention*. I find it is very difficult to accept that the Medicean satellites did not exist before Galileo Galilei 'invented' them, or that the intermediate bosons did not exist before Weinberg, Glashow and Salam 'invented' them.

Therefore, we have either an epistemological (idealistic or realistic) monism crushing knowledge into what is known, or the difficulties of the epistemological dualism. A way out must be found. This may well be possible only by tackling the problem differently, and this is exactly what I am pursuing.

1.3.1 Phenomenological description and transcendental constitution

In the *Prolegomena to Any Future Metaphysics*, Kant reconsiders a classical distinction: the difference between the analytical (or regressive) method, typical of the *Prolegomena*, and the synthetic (or progressive) method, typical of the *Critique* (compare P, § 5). The former starts from the conditioned to arrive at the conditions, the latter starts from the conditions to arrive at the conditioned. Along this line, in the next section, I will proceed analytically, that is, I will ascend from what is conditioned – perception – to its conditions. In the section following that, I will cover the reverse path. However, the analytical ascension must be considered a simple description that has nothing to do with the correctness of the epistemological way. In this case, what appears to be phenomenologically first (that is, perception) is not epistemologically first, since representations are epistemologically first.

The phenomenological description

From a phenomenological point of view, perceiving something is perceiving an indistinct bundle of a multiplicity of features. However, according to my epistemological presupposition, these features are due, in a cognitive constitutive and not causal sense, both to the representations applied and, needless to say, to the perceiving subject's physio-psychological state. Such an indistinct bundle, that is, the *perceived object*, is the brute fact on which the knowing subject has not yet exercised his capability of intersubjectivization, that is, his capability of separating what the concern of all subjects' perception is from what the concern of his own perception is. It could also be called *phenomenon*, even if not in the Kantian sense of *Erscheinung* – that is, as objective appearance – but in the Brentanian sense of *Phänomenon* – that is, as something which presents itself to the mind exactly as it presents itself. Nevertheless, and besides its name, such a perceived object is not yet ready to be intersubjectively known, it is still a sort of bolus which must be digested, that is, it must be intersubjectivized by emending the subjectivity of perception. Therefore, to arrive at an intersubjective object, we must depart from the perceived object by modifying it.

Up to now, we have the *brute fact embedded in subjectivity*. However, and this should be borne in mind, it – being a fact (from the Latin *facere*, to make) – *is something already made*, in the sense that it *has been made cognitively significant by the representations*. If perception were theoretically neutral, human understanding would be an *intellectus ectypus*,[32] but this is not so. Moreover, talking about the theoreticity of perception does not mean talking about the complete production of what is perceived: human understanding is not even an *intellectus archetypus*. *The perceiving subject does not construct reality at all, but selects, makes significant, that is, the subject cognitively constitutes (by means of the applied representations) certain aspects of reality he/she subjectivizes (by means of the individual physio-psychological state), and then objectivizes (by eliminating what pertains to subjectivity).*

Thus, the perceived object is quite different from what really is – whatever this 'what really is' can be, beyond its being the necessary, but obscure and undetermined, background of perception and, therefore, of knowledge. Only our representations and our physio-psychological states can determine aspects (*only aspects*) – by constituting them – of metaphysical reality, which, as such, we know neither intentionally nor empirically. 'What really is', that is, metaphysical reality, is there: mute and waiting for those of us who unveil it by cognitively representing some of its aspects.

Perception is never the perception of reality in its metaphysical purity, but always the perception of the partiality permitted by our physio-psychological states and our representations. Therefore it is possible to perceive only within the limits of both the human perceiving system (even if amplified by suitable instruments), and the given applicable set of representations. Without a telescope and without suitable representations, that perceived object cannot be made cognitively significant as 'mountains of the Moon'. Without a particle accelerator and without the suitable representations, that perceived object cannot be made cognitively significant as 'tracks of elementary particles in a bubble chamber'. This does not mean that the mountains of the Moon or the tracks in a bubble chamber do not exist, but that we are not capable of perceiving that something as 'mountains of the Moon', or as 'tracks of elementary particles in a bubble chamber', without the suitable amplification of the physiological apparatus, and without the suitable representations. Nevertheless, we have here two different impossibilities. The first is a *physical impossibility*; the second is a *semantical impossibility*. The elementary particles are not physically perceivable in any way without a particle accelerator; they are not semantically perceivable as 'tracks of elementary particles' if we do not have the suitable representations. Note, however, that they could be perceived as something else, for example as 'paths on a monitor', if we have the suitable representations to make cognitively significant things such as 'paths', 'monitor', 'paths on a monitor'.

This enables us to understand how the distinction between '*to perceive* something' and '*to perceive that* something is something' can be misleading. *It is impossible to perceive something without that something is something*. Yet, it is possible that one subject perceives something as A, and that another subject perceives *almost the same* something as B. It depends on the applied and applicable representations. It is possible to perceive something as 'tracks of elementary particles in a bubble chamber' if the suitable representation has been applied. However, it is possible to perceive *almost the same* something as 'paths in a monitor', if the suitable representations have been applied. *Perception is always perception that something is something*. And what permits us to make that something perceived cognitively significant is the set of the representations possessed by the perceiving subject.

Nevertheless, it should not be forgotten that the perceiving subject's particular physio-psychological state characterizes the perceived object in a peculiar way. Given the uniqueness of the subject, the perceived object is unique and irremediably connected with it. Therefore, a problem arises: how can a subjectively perceived object be transformed into an objective object? Let us consider a book. It has shape, dimension, colour and so on, and it is perceived with a certain shape, certain dimensions, certain colours, from a certain perspective and so on. If one looks at it from a different point of view, the shape, the perspective, the dimensions, the colours, subjectively change. Nevertheless, we are aware that something remains invariant. *This 'invariant something' is the product of the intersubjectivization of subjective perception and the invariant is the intersubjective.* In this way, the subjective is eliminated and becomes a sort of vague background: everyone knows that it exists, but everyone knows that it cannot be communicated: were one to try, it would be objectivized and so emended.

The intersubjective is what remains invariant as far as the various individual modes of perception on the part of the same subject are concerned, but it is also what remains invariant as far as the individual modes of perception of various subjects are concerned. Each subject looks at the book in his/her own way: from his/her own spatial perspective, with a certain shape, certain dimensions, certain colours. Nevertheless, there is something that all the subjects look at in the same way. There is something invariant: its intersubjective shape, its intersubjective dimension, and its intersubjective position. What remains is what all the perceiving subjects share. Therefore, it is the intersubjectively perceived object that transcends individual modes of perception, and remains invariant in the space-temporal flux of the single modes of perception of the same subject, or of many subjects. In the individual perception, that particular tone of colour I see is only mine; that particular tactile sensation is only mine; that particular fragrance I smell is only mine; that particular intensity of noise I hear is only mine; that particular perspective from which I look is only mine. But as soon as the *perceived object* is intersubjectivized, turning in this way into an *intersubjectively perceived object*, all that belongs to the individual subject is emended and all that belongs to all perceiving subjects remains: that tone belongs to everybody, that roughness of the surface belongs to everybody, that fragrance belongs to everybody, that perspective belongs to everybody. (*Of course, all belongs only to those who want to, and semantically can, perceive it.*)

However, the invariance of intersubjective perception is not sufficient to remark the genuine existence of what is perceived. Since the intersubjective object is invariant for intersubjective perception, the possibility of self-illusion should be avoided. Nevertheless, someone could raise an objection: 'In this way, the self-illusion of a single perceiving subject is avoided, but not the possibility that all the perceiving subjects are victims of the same illusion.'

This is a useful objection, which makes us understand that not everything which is intersubjectively perceived has the same status. Some of the intersubjectively perceived objects are mere collective illusions, and it would be wrong to consider them simplistically existent. To prove the genuine existence of an object neither the invariance for the single subject, nor the invariance for many subjects are sufficient. *There must be invariance for all subjects who want to perceive it here and now, or in whatever future they wish. The genuine existence of an object must be at everybody's disposal.*

A subject can affirm he or she perceives a goblin. However, this perception is a self-illusion if it is not invariant for many subjects. Nevertheless, it might happen that many subjects affirm they perceive a goblin and, in this case, the perception would be intersubjectively invariant. But, again, this invariance is an illusion if it is not empirically detectable by whoever wants to detect it. Whoever affirms perception of some invariance must accept that any other subject can detect it as well. This means that the intersubjective invariance of the intersubjectivized perceived can be detected by whoever wants to test it. Indeed, whoever affirms that they perceive something intersubjectively invariant should provide the method by which empirically to detect such intersubjectivity. Whoever affirms that they perceive something intersubjectively invariant must accept that any other subject can empirically detect such invariance.

At this point, we can call *observable* that object which is not only intersubjectively perceivable here and now, but which is also perceivable by whoever wants to perceive it in whatever time. That is, an observable object is a perceivable object whose intersubjective invariance can be empirically detected by anyone. It is precisely observability, defined in such a way, which warrants existence. Therefore, *observables* or *empirical objects* are the same thing (even if from different points of view). This means that *even if each empirical object is an intersubjectively perceivable object, not every intersubjectively perceivable object is an empirical object. To be such its invariant features must be at anyone's disposal.* Therefore, *observing something* means both (1) limiting our perception to what is invariant (and thus emending what belongs to us as individual perceiving subjects), and (2) being aware that what is observed (an invariance) must be detectable by whoever wants to detect it.

It should be noted that there are objects, for example a celestial body before a gravitational collapse, which can be perceived only once. However, this does not mean that they are not observable. For they can be perceived by more than one subject and, moreover, they can be recorded, and the record can be perceived and detected (that is, observed) by whoever wishes to do it. We are no longer in the case of the synchronic observation of the (at this point historicized) object, but in the case of its diachronic observation, based on its record fixed in physical archives consultable and detectable by all.

Note that the univocity of the observability of the empirical objects existing only in a determined time interval is totally different from the univocity of individual perception. The first is a univocity perceived, or perceivable, by more than one subject. Moreover, it is a univocity which is recordable, that is, testable also beyond the time interval of real existence. The second is a univocity due to the particular physio-psychological states of the individual perceiving subject.

The transcendental constitution

In the section above, I started phenomenologically from the subjectively perceived object to reach the intersubjectively perceived object, that is, the empirical object, so long as its intersubjectivity is also empirically detectable by whoever wishes it. Now, I will analyse the same matter from a transcendental point of view, that is, by starting from the conditions of knowableness: the representations.

We know that concepts have a natural objectivity, since their sense is graspable by whoever possesses the suitable semanticizing area. Thus, it is an objectivity which should be intended as a *semantical objectivity*.

I have also suggested three levels: the representational level, the intentional level, and the empirical level. Moreover, I have argued that the sense determines both the intentional and the empirical reference. Therefore, if the sense of a concept is semantically objective, so is its reference. It follows that *each intentional and empirical object is semantically objective.*

The same result can be reached in another way. Since concepts are semantically objective and since among them there are the property-concepts, it follows that these too are semantically objective. Hence, their references, that is, the intentional properties, are semantically objective. By taking into account that each object is a knot of properties, again we have that *each intentional object is semantically objective.*

While both the intentional and the empirical object are semantically objective, only the empirical object is *empirically objective.* For, as we have seen, the empirical object is characterized by that form of objectivity correlated with its intersubjective and detectable invariance, that is, with the fact that it is an observable.

Hence, even if the empirical object is constituted by the correlated object-concept (that is, a representation), it, as a knot of invariant empirical properties, must be intersubjectively detectable. Therefore, *the empirical object is objective from two points of view*:

(1) *it is empirically objective,* as it is a knot of invariant empirical properties, which are detectable by all the subjects who want it;
(2) *it is semantically objective,* as it is the empirical reference of a given concept, which is a semantically invariant knot of property-concepts graspable by whoever possesses the apt semanticizing area.

While the empirical object is objective both in the empirical sense and in the semantic sense, the intentional object is objective only in the semantic sense. For the latter can be considered only from the point of view of the intentional properties concurring to form it as an intentional knot. Therefore, the empirical object can be thought of as an interface between the phenomenological approach, which regards it as a detectable invariant, and the transcendental approach, which regards it as an invariant semantically constituted. Hence:

(1) as there is independence of the epistemological essence from existence, so there is independence of the objective in the semantic sense from the objective in the empirical, and therefore detectable, sense;
(2) as the domain of the epistemological essence is greater than that of existence (*all that exists has an epistemological essence, but vice versa does not hold*), so the domain of the semantic objectivity is greater than the domain of the empirical objectivity (*all that is empirically objective is semantically objective, but vice versa does not hold*).

It follows that the existence of an object is not implied by its semantic objectivity, but by its observability, that is, by the detectability of its empirical invariance. I do not know that a chair exists because it is semantically objective, but because it is observable. I do not know that an electron exists because it is a knot of invariant intentional properties, but because it is a knot of invariant empirical properties, that is, because it is observable. I know that Pegasus does not exist because it is not observable, even if it is a knot of intentional invariant properties.[33] *Only the detectability of the perceivability of the intersubjective invariance, that is, only observability, warrants existence.*

I have pointed out that the empirical object is never detected as a whole. What is detected is a given empirical property. Therefore, the notion of the detectability of an empirical object as a whole must be clarified. In fact, we should refer to the intersubjective detectability of the individual properties concurring to its determination. We can never observe the electron as a whole, but its empirical properties one by one. We observe that it has a given charge, a given mass, a given spin, and so on. Now, since the empirical object is thought of as a knot of empirical properties made cognitively significant by the semanticizing area, it must not be thought of as a piece of metaphysical reality, but as a bunch of aspects of metaphysical reality constituted in that particular way. In other words, *each individual empirical property concurring in determining the empirical object highlights an individual aspect of reality, that is, that there is something capable of interacting with our experimental apparatus in a way made cognitively significant by the semanticizing area*. Thus, what we intersubjectively perceive is an individual aspect of metaphysical reality (whatever this latter may be) constituted in a certain way. This is what urges me to argue that *each object*:

(1) from the point of view of what cognitively constitutes it, is *a knot of intentional properties*;
(2) from the point of view of its observability, is *a knot of empirical properties*;
(3) from the point of view of what is cognitively constituted, is *a knot of aspects of metaphysical reality*.

Therefore, we produce object-concepts by producing knots of property-concepts. In so doing, we automatically produce the intentional references of the object-concepts, that is, the intentional objects, and the intentional references of the property-concepts, that is, the intentional properties. If we empirically observe such properties, thus they are also empirical properties. In this case, we have also constituted in a certain way certain aspects of otherwise metaphysically undetermined reality.

Earlier we arrived at the conclusion that what is epistemologically first, as far as the constitution of the empirical object is concerned, is the property-concept, and, as far as the sense of the concept is concerned, are the representations, that is, the semanticizing area to which the concept belongs. As we have seen, changing the semanticizing area means changing the sense of the concepts, and thus the cognitive constitution of the empirical objects.

By producing the representation called 'Lorentz's electrodynamics', the intentional 'electron' is produced. This is a knot of particular intentional properties determined by that representation. In this case, there is also an empirical 'electron'. Therefore, the empirical object called 'electron' is thinkable of as a detectable invariant, constituted in that way by Lorentz's representation. Each individual empirical property concurring to form it as a knot can be independently detected by means of a suitable experiment. The same happens in the case of the electron of Quantum Electro Dynamics (QED). There is the concept of 'electron'. This has an intentional reference and an empirical reference which are, respectively, a knot of intentional properties and a knot of empirical properties indicated by the QED. Since the semanticizing areas of the concepts 'Lorentz's electron' and 'QED's electrons' are different, the respective intentional and empirical references are also different. That is, we have two different observables, two different knots of empirical properties, two different knots of aspects of reality. However, it would be wrong to affirm that they indicate two different metaphysical realities. We do not know, either intentionally or empirically, if metaphysical reality is the same or not. We know only that the concepts 'Lorentz's electron' and 'QED's electron' have permitted us to make cognitively significant different aspects. Moreover, of course, this does not imply in any way that there is a void intersection between the property-concepts concurring in determining the Lorentz's electron and the property-concepts concurring in determining QED's electron.

We *produce* conceptual representations, by means of which we *cognitively constitute* (and therefore *we do not construe or invent*!) aspects of reality. Nevertheless, we produce representations which must be thought of as conjectures; therefore the existence of empirical objects, and thus of the correlated empirical properties, is also a matter of conjecture. But if we empirically detect empirical properties, we constitute previously neglected aspects of reality. This is the reason why investigating reality does not mean inventing reality, or, worse, constructing reality, even if knowledge (epistemologically but not phenomenologically) begins with representations. The representations produced by the knowing subject tell us what the knots of empirical properties, and thus the knots of the aspects of metaphysical reality, could be.

Note that, implicitly, I am discussing the possibility of enlarging knowledge by enlarging the number and the power of the representations. This can only be achieved by hypothetically constituting new aspects of reality. Producing representations, for example doing theoretical physics, means producing intentional references by means of which we could account for (and thus constitute) not yet unveiled aspects of metaphysical reality. Working with observables, for example doing experimental physics, means testing if those representations have empirical references, that is, testing whether they are empirically detectable. A test of the empirical detectability of intentional properties, and thus a check on the hypothesized empirical properties, becomes a test of the cognitive validity of the representations: a test of their capacity of being representations of empirical objects.

With this approach, the problem of the absolute truth or the absolute falsity of a representation loses its importance. What is important is its cognitive validity as representative of certain aspects of the world. Classical mechanics makes certain aspects of the world significant; this is why it is a valid scientific representation. Equally valid is special relativity, which makes other aspects of reality significant. Of course, such validities are confined within certain limits: each representation of the world is valid within limits which can be empirically determined. When, by means of an experiment, the observability of a property is tested, implicitly the limits of validity of the representation to which it belongs are investigated.

When, in 1916, Millikan carried out the experiment to test whether certain properties, constituted as such by the Einsteinian representation of the photoelectric effect, existed, he tested not only the representational power of Einstein's proposal, but he also highlighted the limits of the validity domain of classical electromagnetism. Moreover, that what had been hypothesized by Einstein was observed does not at all imply that his theory must be considered absolutely true. Instead, it simply means that what was intentionally proposed had also an empirical counterpart. There is another reason why this historical case is instructive. Between 1886 and 1887, Hertz, during his experiments to detect the effective observability of the Maxwellian equivalence between electromagnetic waves and light, noted something strange. This 'something strange' was the photoelectric effect. But Hertz had no representation permitting him to make it a cognitively significant aspect of the world. The result was that it remained undetermined.

This story can be used as a further argument, if still necessary, that the theorist does not invent reality, but produces representations which can make otherwise neglected aspects of the world cognitively significant. The photoelectric effect was *not invented* by Einstein, but it was *discovered* by Einstein's proposing a representation that made it cognitively significant. Einstein *unveiled* aspects of reality otherwise relegated to metaphysical undetermination.

However, it is inevitable that sometimes we recognize that certain representations are totally inadequate to grasp aspects of reality. In this case, experiments shed light on the fact that representations either do not make any aspect of reality significant, or that what they make significant can be done better by other representations. Those representations can be shelved. But setting them aside does not mean that the intentional objects that the representations refer to must no longer be considered objective. They are always knots of intentional properties, even if these have no empirical counterpart and, therefore, there is no point in maintaining the correlated representations in the game of knowledge. As a result, the non-valid (not the false) representations enter a sort of limbo, from which they may emerge at another point in time.

This was the story of the representation of the 'plum cake' atom proposed by Thomson. Around 1910, Rutherford and his colleagues analysed the results of the experiments of the scattering of α particle from a metallic target. They arrived at the conclusion that the positive charge of the atoms had to be confined in a very small space. This revealed the total inadequacy of Thomson's representation. Therefore, Rutherford's experiment, by pointing out the detectability of the hypothetical aspects of the world proposed by Thomson's representation, had two consequences: (1) it spurred the search for a new representation capable of making fully significant the recently experimentally unveiled aspects of reality and (2) it caused Thomson's representation to be shelved.

To conclude, each representation is valid in its validity domain, which can be systematically determined by means of experiment. However, each representation constitutes knots of aspects that cannot be constituted in the same way by the other theories.

1.3.2 Unobservability and observability by inference

Since it is the observability of an object that permits us to speak about its empiricity, and therefore about its existence, a representation that wants to grasp the world must be produced so that the objects it deals with are observable. Therefore, from the point of view of knowledge of the world, those representations proposing objects that are unobservable in principle, that is, objects which can be considered only at the intentional level, are open to criticism.

Such objects may be introduced to give formal consistency to a theory; at other times, they are the heritage of previous (often pre-scientific) conceptions. However, their presence induces us to suspect that the representation containing them could either be replaced with another one not involving their introduction, or could be maintained without their presence.

This *theoretical unobservability*, exemplified by a huge number of historical cases, is a form of unobservability strictly connected with the representations themselves. It happens that a given object has unobservable properties, that is, it does not have empirical properties but only intentional properties, owing to the way in which the representation has been produced. Let us consider aether. Towards the end of the nineteenth century there were two schools: the first attributed mechanical properties such as density and elasticity to the aether; the second, emblematically represented by Lorentz (1909), denied this form of mechanistic reductionism. Consequently, according to the former, aether was observable since it had potentially empirical properties. According to the latter, aether was theoretically unobservable since – as Lorentz suggested – it was the same inside and outside matter and without any causal properties (Ch. I, § 7; Ch. V, § 194). The representation – that is, Lorentz's theory of electron – was produced so as to admit a purely intentional object. However, this induced the suspicion that it was possible either to replace the theory with a new one (that is, special relativity) where this object was no longer possible, or to use it in a domain where such an object was negligible (classical electromagnetism emended of aether continued to be used in given domains).

Besides theoretical unobservability, *technological unobservability* must also be considered. This form of unobservability is due not to theoretical constraints but to technological constraints. For example, to render mathematically coherent the standard model of elementary particles, beyond the four known fields (gravitational, electromagnetic, weak and strong fields), a new field has been introduced: Higgs field (from the name of the physicist who proposed it: Peter Higgs). Higgs field is a scalar constant field and it is thought of as the cause of the mass of all particles. By interacting with Higgs field, particles 'absorb' energy, and because of the equivalence between mass and energy, they acquire mass. As a consequence, this field cannot be universal, since it interacts differently with different particles (for example, it does not interact at all with massless particles, since if they interacted they would 'absorb' mass). However, the introduction of Higgs field obviates both the purely physical problem of the mass of particles, and the mathematical problem of the coherence of the standard model. Yet, does it exist?[34] To answer this question positively, the particle that mediates the interaction with it – that is, Higgs boson – should be detected. However, this is not an easy task, since at present there is no particle accelerator capable of reaching sufficient energy. In our language, we have a representation – the standard model – which, among its intentional references admits a particle: Higgs boson. The standard model becomes a good representation of the world if the intentional Higgs boson is also empirical, that is, if a suitable experimental apparatus can be constructed and if the experiment gives a positive result. However, up to now, no suitable experimental apparatus can be constructed owing to the technological limits. It means that, up to now, Higgs boson is technologically unobservable.

In these cases, negative judgement on the scientific representation is not so simple and immediate. For many formerly technologically unobservable objects have became observable through technological development, as happened in the case of gravitational lenses. Nevertheless, others are still unobservable. There is no unambiguous demarcation criterion between what is technologically unobservable today but tomorrow will become observable, and what is technologically unobservable both today and tomorrow. The valuation of the individual case must be remitted to the individual scientist, or scientific community. Only scientists, or scientific communities, can judge the validity of a theory with technological unobservables (and, of course, they can also arrive at a wrong result). These may always remain unobservable or, if there is a suitable technological development, it may be possible to arrive at the conclusion that they do not exist at all.

It is experimentation that permits us to test the effective existence of an otherwise purely intentional object. Only a positive experimental result can be proof that some intersubjective perception has happened.

At this point the question concerning the fact that many empirical objects can be observed only indirectly by means of the observability of their effects comes into play. The observability of a Z^0 particle is different from the observability of a chair, or of the mountains of the Moon. The difference is so great that many consider the Z^0 not only as an unobservable particle, but also as a non-empirical particle. Although

I agree that there is a difference between the two kinds of observability, I do not agree on the unobservability of the Z^0, and still less on its non-empiricity. Instead, I argue the observability of the Z^0, even if in this case we must describe it as *observability by inference*. In this case, the object is observed not only by means of an experimental apparatus, but also by means of a causal representation. This enables us to infer the indirect observable cause from the direct observation of its effects. We can infer fire by observing smoke and by means of a causal representation, which is testable independently of the specific case. Analogously, we can infer the Z^0 by observing its tracks and by means of a causal representation, which is testable independently of the specific case. In both cases, the fire and the Z^0 are observable by inference. It is true that we can approach the smoke and observe the fire directly, whereas we cannot do this in the case of the Z^0. However, if we accepted that only what is directly observable is an aspect of reality, we would know practically all there is to know about the world. Is this a good position to argue? It should also be remembered that viruses were observable only by inference until the 1930s. Fortunately, biologists, from Pasteur onwards, considered them as empirical objects. This teaches us that *what is observable by inference today can become directly observable tomorrow*.

Not all authors admit the existence of observables by inference. For example, van Fraassen (1980) – who declares himself a constructive empiricist – 'suspends judgement' on their reality. Yet, where is the validity of this position? Let us look at how a particle detector works, in particular let us consider the bubble chamber.

A container holds overpressurized liquid (often hydrogen or a neon-hydrogen mixture, or propane). This is immersed in a magnetic field that deflects the trajectories of the charged particles entering the bubble chamber. Cameras record what happens in the chamber. When the incident beam enters the chamber, there is a diminution of the pressure due to the expansion of a piston. The diminution of pressure and the ionization of the atoms of the liquid, due to the incident charged particles, cause each ionized atom to become a nucleation centre of bubbles. Bubbles grow along the track of a particle. The bubble track (deflected by the magnetic field) is photographed. Thus, in the resulting film there is the *macroscopic* record of the trajectory of the *microscopic* elementary particle. In this case, the elementary particle (an electron, a positron, a muon, or whatever is charged and microscopic) is not directly observed. Instead, it is observed by inferring from its effects in a bubble chamber, that is, by inferring from a bubbled deflected trajectory. By analysing the geometric characteristics of the track (that is, the effect) what kind of particles they are (that is, the cause) is inferred. However, to do this causal representations are necessary.

Just a short reflection enables us to comprehend that in this case at least two causal theories are needed: one which accounts for the generation of bubbles in an overpressurized and superheated liquid, and another which accounts for the interaction between charged particles and matter. The former is phenomenological thermodynamics, the latter is classical electromagnetism. These are the causal theories which allow us to infer the existence of a directly unobservable object by means of the observation of the existence of a directly observable object. If we really wanted to deny, or to suspend judgement on, the epistemological (not *metaphysical*) reality of elementary particles by adducing the fact that they are not directly observable, we should also deny the fact that phenomenological thermodynamics and classical electromagnetism are empirically adequate in their validity domain. In all honesty, this seems a strange position to hold.

As in the case of the bubble chamber, the representations which allow us to infer observable effects from non-directly observable causes are representations which have been corroborated in an enormous number of cases – totally independently of the case given here concerning the detection of microscopic charged particles. Denying their epistemological reality would mean either (1) also denying these causal representations in their validity domain (which would be rather hazardous), or (2) behaving in a schizophrenic way: accepting them in all cases except those in which they are used as a connection between the directly observable and the observable by inference (an intriguing but bizarre solution).

1.3.3 A short conclusion

I have argued that the intentional object is a knot of intentional properties produced by the representation to which it belongs. Moreover, I have shown that some object-concepts also have an empirical reference, besides their intentional reference. In this case, the object is observable. That is, it is an empirical object.

In this way, I have not reduced the empirical object to the mere appearance of an unknowable noumenic reality. Instead, the empirical object is intended to be a knot of properties constituting in a cognitively significant way some aspects of metaphysical reality, on whose extension and structure, nevertheless, I suspend judgement. Note, however, that *this proposal does not at all involve the dissolution of the world into the representations*. I have limited myself to claiming that we, knowing subjects, produce representations, which admit intentional references (the intentional objects) and, in given case, empirical references (the empirical objects). If we have representations which admit empirical references, the former constitute, in a cognitively significant way, the world. From this point of view, criticism of ideas of indeterminism or determinism of metaphysical reality, or about its causality or acausality, its being chaotic or ordered, becomes an easy task. These concern only the way in which reality has been constituted, and not how reality is in itself. It is a fallacy to claim that metaphysical reality is indeterministic or deterministic, causal or acausal, chaotic or ordered starting from what a given representation says. We do not know anything about metaphysical reality beyond what has been constituted by representations admitting empirical references. Therefore, we should limit ourselves to affirming that certain representations, produced by the knowing subject so that to be indeterministic or deterministic, causal or acausal, chaotic or ordered representations, constitute, and therefore make cognitively significant, certain aspects of the world as indeterministic or deterministic, causal or acausal, chaotic or ordered aspects. Inferring that metaphysical reality is deterministic or indeterministic, causal or acausal, chaotic or ordered only because a handful of its aspects have been made significant in this way is, as said, a fallacy: the fallacy of the *transitus de genere ad genus*: we give at a certain element of a class (reality) a property (to be deterministic or indeterministic, causal or acausal, chaotic or ordered, and so on) that, instead, belongs to another class (representations).

The knowing subject does not know how reality actually is; neither does the subject construct reality; instead he or she cognitively constitutes it by means of representations (and physio-psychological states). Therefore, my wise realism involves neither the idea that reality is dissolved into representations, nor the idea that representations mirror metaphysical reality. In such a way, the Scylla of radical constructivism and the Charybdis of naive realism are avoided.

2
Laws of Nature

In *The Structure of Science,* Nagel is extremely pessimistic about the possibility of explicating in a rigorous way the concept of 'law of nature':

> The term 'law of nature' is undoubtedly vague. In consequence, any explication of its meaning which proposes a sharp demarcation between lawlike and non-lawlike statements is bound to be arbitrary. There is therefore more than an appearance of futility in the recurring attempts to define with great logical precision what is a law of nature – [by] virtue of its possessing an inherent 'essence' which the definition must articulate. For not only is the term 'law' vague in its current usage, but its historical meaning has undergone many changes. We are certainly free to designate as a law of nature any statement we please [...] It would indeed be futile to attempt an ironclad and rigorously exclusive definition of 'natural law'. It is not unreasonable to indicate some of the more prominent grounds upon which a numerous class of statements is commonly assigned a special status.
>
> (Nagel, 1961, pp. 49–50)

If we followed Nagel's line on this topic, there would be little point in trying to work on it. In fact, Nagel's graphic and discouraging judgement can be seen as the honest conclusion at which a good observer of the discussions on laws among the German and Austrian neopositivists and the American post-positivists must arrive.

However, after Nagel's dismal epigraph on the failure of the neopositivist and post-positivist attempts to grasp nomologicity, something strange happened. While the neopositivist and the post-positivist attempts were strongly characterized by an emphasis on the logical structure of the statements supposed to be laws, and on a radical refusal of both modal approaches and metaphysical solutions, since the 1960s both the modal and the metaphysical way have re-emerged, as if all the previous doubts have been forgotten. This passage from a strong anti-metaphysical position to a strong metaphysical one was enabled by a kind of interregnum during which supporters of the counter-factual approach provided the link. Needless to say, the metaphysical attempts have been severely criticized, also because the memory of the anti-metaphysical heritage was not completely vanished. Nevertheless something apparently stranger happened.

As is known, the neopositivist and the post-positivist positions had a strongly Humean flavour. Hume was one of the noble fathers of German-speaking neopositivism and he retained this position during the period of American post-positivism, that is, during the so-called 'standard view' period. And, as is also known, Hume was an outspoken opponent of metaphysical ideas:

> When we run over libraries [...] what havoc must we make? If we take in our hand any volume; of divinity or school metaphysics, for instance; let us ask, Does it contain any abstract reasoning concerning quantity or number? No. Does it contain any experimental reasoning concerning matter of fact and existence? No. Commit it then to the flames: For it can contain nothing but sophystry and illusion.
> (Hume, 1748, p. 135)

Unfortunately, the new metaphysicians had forgotten both Hume and their coeval Humeans, so they tried again to solve the problem of what a law is by means of essences, intrinsic necessities of nature, universals *de re* and so on. This continued until van Fraassen appeared on the battlefield. Brandishing the sabre of his counter-argumentative ability, in his *Laws and Symmetry*, he launched a strong criticism of the new metaphysicians of laws and accused them of being pre-Kantians.

Never had an epithet seemed more correct, although something did not add up. We know that Kant, in his *Prolegomena to Any Future Metaphysics*, writes clearly that Hume awakened him from his 'dogmatic slumber'.[1] We also know that many of Kant's coevals, in particular, Mendelssohn, nicknamed him the *Alleszermalmender*, since he was considered as the one who had destroyed all the beliefs of the validity of metaphysical certainties, as a sort of 'demolition man' of classical metaphysics. In the same way we could consider van Fraassen as the *Alleszermalmender* of contemporary metaphysical positions on laws. But (there is always a 'but') even if Kant was woken up by Hume and even if he was called the *Alleszermalmender*, beyond a *pars destruens*, he proposed also a *pars construens*. That is, Kant understood that it was impossible to be metaphysicians in the pre-Humean way, and proposed his own philosophy. As far as neopositivism and post-positivism are concerned, things went differently. They were anti-metaphysical, in a strongly Humean way, and part of their role was to wake many philosophers from their own dogmatic sleep. But as soon as these two traditions showed their philosophical weaknesses and were abandoned, the metaphysical temptations rose again, abandoning as well the Kantian alternative. Then van Fraassen arrived and accused them of being pre-Kantian. But van Fraassen himself did not move along a Kantian path. In his writings, Kant is a kind of 'stone guest', like the Commendatore in Mozart's *Don Giovanni*. In fact, it would have been sufficient that van Fraassen claimed that the new metaphysicians were pre-Humean. If both the neopositivist and post-positivist proposals were Humean, the new metaphysical proposals rather than being pre-Kantian, must be considered as pre-Humean. They are a step backward rather than a step forward. Moreover, van Fraassen's criticisms appear to be more Humean than Kantian.

But, why not challenge us with the Kantian line of thought? In fact, only a handful of philosophers are moving along this path. This seems bizarre, not least because an increasing number of philosophers claim that there are no laws or that laws are mere models that have nothing of relevance to say on the world. However, they make this claim without having investigated all the theoretical possibilities offered by our (and their) historical heritage. In what follows, first I will show the failure of both the Humean-like and the successive pre-Humean-like attempts. In doing so I will argue that they do not solve the problems raised by Schlick, who I consider as the forerunner of this way of thinking. Then, since it is obviously unacceptable to consider Kant only as a 'stone guest', I will reconstruct his view on laws. Finally I will propose my own Kantian solution by emphasizing the role of the idea of *unity of system*.

However, at the start, a terminological note is worthwhile. This is important in order to grasp both what was discussed during the neopositivist and post-positivist ages, and what Kant was discussing. In this way, besides clearly delineating the problems really at issue, I have the opportunity of paving the way for my proposal. The note concerns the difference between three terms: (1) lawlikeness; (2) lawness; (3) lawfulness. *Lawlikeness* concerns certain statements, in particular the universal conditionals, that have the form of a law, even if it is not said that they are laws. *Lawness*[2] has to do with the necessary and sufficient conditions characterizing a law. Therefore lawness is what makes a lawlike statement a law. What I will call 'Schlick's problem' is precisely the problem concerning lawness, that is, in other words, nomologicity. Finally *lawfulness* will be used to indicate, with Pearson (1892, p. 72), 'what is not prohibited by the law'. Thus I will speak of the lawfulness of nature to indicate what is allowed and not allowed in *nature*.[3] Kant discusses both lawness and lawfulness. Nevertheless this difference has not been well thematized in contemporary discussions on laws, since (1) for the regularist, *nature*[4] is not lawful and what is important is only the lawness of lawlike statements; (2) for the metaphysicians, *nature*[5] is of course lawful, and the laws grasp it necessarily; therefore analysing lawness means implicitly analysing lawfulness and there is no point in discussing the two notions independently. On the other hand, for a Kantian approach, it is extremely important to distinguish between the problem concerning lawness and the problem regarding lawfulness,[6] and I will try to offer a solution to both of them.

2.1 The failure of the Humean and pre-Humean attempts

2.1.1 Schlick: the dawn of the contemporary debate

One of the purposes, if not the main one, of the philosophers belonging to the composite neopositivist movement regarded the search for a rigorous criterion demarcating not cognitively significant statements from cognitively significant statements, in particular in relation to scientific statements. As is known, it was thought that such a criterion could be identifiable with empirical verification. Immediately a problem arose: the laws of nature are universal statements and therefore not verifiable in a conclusive way; what then, should we do?

Among the first presentations and analyses of this problem, the most striking, for its theoretical clarity and far-sightedness, was that proposed by Schlick in his 1931 'Causality in Contemporary Physics'. His discussion starts with the following observation: 'for whatever the distribution of the given [empirical] quantities may be, it is *always* possible, notoriously, to find [mathematical] functions which depict precisely this distribution with any exactness we please' (Schlick, 1931, p. 181). This assertion implies two things:

(1) any set of observed empirical data can be represented with the precision we want by a mathematical function;

(2) the same set of observed empirical data can be represented with the precision we want by more than one mathematical function.

In both cases, the second of which concerns the familiar topic of the underdetermination of mathematical representations by data (that is, the underdetermination of theories by data), we need to distinguish between 'good' mathematical functions, that is, the mathematical representations of nomological regularities, and 'not good' mathematical functions, that is, the mathematical representations of accidental regularities.

To solve this problem, first Schlick investigated the possibility of imposing what was known as *Maxwell's requisite*, according to which values of space-time coordinates must not appear in the laws of nature (see Maxwell, 1873, II). This led to the identification of a further problem:

> [...] would it be a necessary condition? This we shall scarcely be able to say, since it is certainly to think of a world in which all events would have to be reproduced by formulae in which space and time explicitly occur, without then denying that these formulae present correct laws, and that this world would be completely ordered. So far as I can see, it would be imaginable, for example, that regular measurements of the elementary quantum of electricity (electron charge) would yield values for this quantity that fluctuate up and down quite uniformly by 5%, in say 7 hours, and then another 7 hours, and then 10 hours, without our being able to find even the slightest 'cause' for this; and perhaps there would be another fluctuation on the top of this, for which an absolute change of the earth's position in space would be held responsible. The Maxwellian requisite would then no longer be satisfied.
>
> (Schlick, 1931, pp. 181–2)

A second requisite, Schlick suggested, could be that of simplicity: the simpler the mathematical expression, the more it nomologically grasps the regularity. Nevertheless, this also proved unsatisfactory, being entirely unclear as to what 'simplicity' should mean.

However, to organize a series of data by a mathematical expression is not sufficient. It is necessary that such an expression permits us to predict new data: 'only if this is so, does he [the physicist] takes his formula to be a law of nature. In other words, the true criterion of [the nomological] regularity, the essential mark of causality, is the fulfillment of predictions' (Ibid., p. 185).[7] This runs up against the fact that the criterion of empirical confirmation, though necessary, also reveals some problems, since the confirmation of a single prediction implies neither the definitive verification of the law – this deals with infinitive cases – nor the certainty that a causal relation really exists in the world (Ibid, p. 51).[8]

At this point, there seem to be two possibilities: either we consider as valid the criterion of verification but we do not interpret the laws as real scientific statements, or we do the opposite. Schlick, however, was inclined to an intermediate solution: on the one hand, laws cannot be considered real scientific statements since they cannot be conclusively verified, but, on the other hand, they cannot simply be expunged as not cognitively significant. The solution he suggested, that he admitted he owed to Wittgenstein (Ibid., p. 188), was based on the idea that even though they cannot be verified definitively, laws 'represent, rather, a prescription for the making of assertions [...since] as we know, it is possible to test only the individual statements that are derived from a law of nature, and these always have the form: "under such and such circumstances, this pointer will indicate that line on the scale", "under such and such conditions, a blackening occurs at this point on the photographic plate", and the like. Every verifiable assertion, and every verification, is of this type' (Ibid., p. 188).

The laws, therefore, are no longer to be considered as real statements, but as schemes to build up singular statements: the only ones that can be conclusively verified.[9] This has to be so, since only when we are faced with a real empirical verification are we – for Schlick – allowed to speak of a causal relationship (Ibid.)

The Schlickean position has two relevant features:

(1) The empirical confirmation of the predictions is considered as a core requisite (I call it *Schlick's criterion*) to demarcate a statement expressing a nomological regularity from a statement expressing an accidental regularity,[10] that is, to use a later terminology, to distinguish a causal nomological statement from an accidental statement (I call it *Schlick's problem*).
(2) It is wholly in concert with the Humean tradition:

> We acknowledge, with Hume, that there is no logical justification for them [the laws]; there cannot be any, because they are simply not genuine propositions. The laws of nature are not (in the logician's terminology) 'general implications', because they

> cannot be verified for *all* cases; they are prescriptions, rather, rules of procedure that direct the scientist to orient himself in reality, to discover true propositions, to expect certain events. It is this expectation, this practical behavior, to which Hume points in the words 'custom' or 'belief'.
>
> (Ibid., p. 197)

Taking all this into account, we see that there are two main reasons for the importance of this 1931 essay for the contemporary setting of the question concerning laws of nature:

(1) First, Schlick completely changed the physiognomy of the problem of laws. While in the pre-neopositivist period, chiefly within the European debate at the end of the nineteenth century, the problem had been tackled, it had been dealt with from an essentially epistemological point of view, that is, discussion focused primarily on the cognitive role of laws. With Schlick the problem divides into two parts and each gives rise to a fruitful research programme:

   (a) The research programme concerning the liberalization of the criterion of significance: if we can consider significant only the statements that can be conclusively verified, the laws of nature, being universal statements, should be considered as non-significant. What then, can be done? As we have seen, in 1931 Schlick proposed an interpretation of laws of nature as schemes of singular statements, the latter conclusively verifiable and therefore cognitively significant.[11] Nevertheless the Schlickean solution did not satisfy everybody, so that this problem, highlighted in 1934 by Lewis (in 'Experience and Meaning'), Popper (in *Logik der Forschung*) and Nagel (in 'Verifiability, Truth and Verification'), was able to lead towards the liberalization carried out in 1934 by Popper himself, and in 1936 by Schlick (in 'Meaning and Verification') and Carnap (in 'Testability and Meaning').[12]

   (b) The research programme concerning the determination of the necessary and sufficient conditions to characterize lawness, that is, lawlike statements as a law of nature (*Schlick's problem*). It should be noticed that before Schlick nobody had ever seriously been interested in fixing the logical or epistemological requisites of nomologicity. After Schlick, the problem entered the agenda of the philosophers of science.

(2) Schlick's work exemplifies the research tradition inside which the neopositivists, the post-positivists of the 'standard view', and the new regularists move. It is a strong Humean tradition which refutes both a causal structure of the world and the necessity, in a strong metaphysical sense, of nomological statements describing the empirical regularities. The problem, for this tradition, is not to understand if or why laws of nature are metaphysically necessary, but to differentiate the nomological regularities from the accidental regularities. Those who accept this approach, that is, the *classical regularists* (for example, Schlick, Reichenbach, Nagel, Pap) and the *new regularists* (for example, van Fraassen),[13] will engage in a programme aiming at finding either logico-classical conditions (for example, Reichenbach, Nagel, Pap), or modal conditions (for example, Burks), or pragmatic conditions (for example, Goodman), or structural conditions (for example, van Fraassen). Those who do not accept this approach, that is, the anti-Humeans (compare the anti-nominalism à la Lewis, the necessitarianism à la Sellars, Pargetter, McCall and Vallentyne; and the realism on the universals à la Kneale, Armstrong, Dretske and Tooley) will take up a research programme aimed at showing that laws are strictly connected to metaphysical nomologicity. It is this duality between Humean regularists and anti-Humean realists which has characterized the main debate on the laws of nature. Let us see, now, these two traditions in detail.

2.1.2 Reichenbach: the search for the explication of the notion of 'law'

If Schlick posed the new problem of the laws of nature and suggested as a demarcation criterion between nomologicity and accidentality the empirical confirmation of the predictions, Reichenbach, in *Elements of Symbolic Logic* (1947, ch. 8) and, above all, in his posthumous *Nomological Statements and Admissible Operations* (1954),[14] attempted a more rigorous approach. His idea, especially in his 1954 work, attempts to explicate, in the sense of Carnap (1950),[15] the notion of 'law of nature' by resorting to the first order predicate calculus with identity. This is essentially why Reichenbach should be seen as the emblematic representative of the 'standard view' of facing Schlick's problem, which reached its climax with Nagel's (1961) great synthesis.

Reichenbach's proposal was not particularly influential and it was largely neglected, perhaps due to the technicality and complexity of its content, although it was quoted by Nagel (1961, p. 64). Moreover, most of Nagel's requirements are nothing but Reichenbach's, even if they are presented in a different and more reader-friendly way, as I show in a while.[16]

In spite of the complexity of Reichenbach's proposal, some aspects must be sketched. His starting point concerns the translation into a formal language of the notion of 'reasonable implication', and therefore that of 'logical implication' and 'physical implication', which he put together in a class called 'nomological implication'. This is the first aspect to take into account. Reichenbach wanted to explicate the nomological implications, a subclass of which is given by the physical implications, that is, the laws of nature. Note that Reichenbach did not consider whether there were laws of nature beyond the physical laws, that is, he did not consider the possibility of biological laws, chemical laws and so on. At that time, the model science was physics, and physics was considered by all the discussants of that age.

Reichenbach did not quote Schlick (1931) but the major problem he was obliged to face to individualize the laws of nature was exactly Schlick's problem (how to demarcate nomological statements from accidental statements?) and the solution he suggested is very close to Schlick's criterion, even if logically reformulated. Anyway, let us move step by step and limit the discussion to one part of Reichenbach's nomological statements, that is, to the laws of nature.

After a long and formal prelude, Reichenbach arrives at the claim that a statement is a law of nature if:

(1) it is verifiably true;
(2) it is an all-statement;
(3) it is universal;
(4) it is unrestrictedly exhaustive;
(5) it is general in self-contained factors.[17]

Let us postpone the first requirement and begin with the others, which are formal requirements. Let us consider a statement written in its irreducible form, that is, stripped of any redundant logical part. According to the second requirement, a law of nature must be an all-statement, that is, a statement containing at least a universal quantifier. In other words, as Schlick also pointed out, a law of nature must be valid for all the cases, that is, it must have an infinite domain of applicability. The third requirement is more problematic: a law of nature must be a universal statement, that is, using Reichenbach's definition, a statement which does not contain 'individual terms'. Correctly, as already identified by Schlick, by 'individual term' Reichenbach means 'a term which is defined with reference to a certain space-time region [... or] a proper name or a definite description' (Reichenbach, 1954, p. 32). Immediately we recognize Maxwell's requisite and we know, thanks to Schlick, that it is neither necessary nor sufficient.

Unlike the requirement to avoid proper names and definite descriptions, it would appear that Reichenbach wanted simply to rule out statements containing elements such as units of measurement, values of universal constants, references to scientific samples and scientific proper names. In fact he was subtler still. He suggested erasing from nomologicity all those statements which do not admit a reconstruction in which the reference to individual terms vanish. Statements containing units of measurement or universal constants can be rewritten so that those particular terms indicating given quantities vanish. For example, the statement 'The speed of light is 3×10^{10} cm/sec' can be rewritten as 'The speed of light is constant', where 3×10^{10} cm/sec is a value that can be cancelled since it is related to a particular standard of measure. The statement 'The charge of the electron is 1.6×10^{-19} C' can be rewritten as 'The charge of the electron is constant', where 1.6×10^{-19} C can be cancelled since it has only practical aims.

With reference to the elimination of the universal constant, Reichenbach writes 'We can *imagine* the system of knowledge as given in the form of the totality of laws of nature supplemented by a table of natural numerical constants. This table is indispensable for practical applications, but does not belong to the nomological part of knowledge' (Ibid., p. 33, my emphasis). Certainly, we can 'imagine' what we want, but I am not really sure that what we can imagine is necessarily good. Are we sure that we can eliminate from science the universal constants? If science is thought of as a useful way of knowing and representing the world we live in, we must consider how this world is; and our world is characterized by particular universal constants. Different universal constants would characterize a world different from ours. Unfortunately, I suppose, in this way our attempt at scientific understanding would shift from science to science fiction. Also the statements containing reference to scientific samples can be rewritten. For example a statement such as 'copper has an atomic weight of 63.5' is a law of nature since it means that 'there exists an adequate rational reconstruction of languages in which this statement is a law of nature', that is, where it can be rewritten without reference to individual terms. And the same holds for statements containing proper names. For example, 'the polar bear is white' can be rewritten as 'each living being having given properties $p_1, \ldots, p_n$ [identifying what, in everyday language, we call "polar bear"] has the properties

$p'_1, \ldots, p'_k$ [identifying what, in everyday language, we call "white"]'. This is true, but are we sure that we have eliminated all the individual terms? Correctly Jobe (1976, pp. 244–5), quoting Smart (1963), points out that when we are speaking of biological entities we must take into account that what we say about them is not valid for each time and for each space but for the space-time portion in which their species, or their varieties, live. Both the $p_1, \ldots, p_n$ and the $p'_1, \ldots, p'_k$ considered above characterize a particular species or variety of beings living in a certain temporal period and in a certain spatial region. In other words, any reference to biological living beings must take into account the evolutionary point of view and any statement about them has to be, implicitly or explicitly, limited to a particular space-time region. Therefore, either we deny any biological statement the status of law, as some would be willing to do, or we should eliminate from the requirements of nomologicity that regarding universality à la Reichenbach.

One could object that Reichenbach was concerned only with physical nomologicity[18] and not with biological nomologicity. This could be true. Nevertheless, what we have said regarding the impossibility that a biological statement holds for each space and for each time is also valid for physical statements. It would be sufficient to recall Schlick's objection to Maxwell's requisite, but since this was based on a bizarre, essentially fictive world, its validity could be questioned. However, we now know that no physical law is valid for any time, so long as we accept the theory of cosmic evolution. Feynman (1965) pointed this out. Before 10^{-43} sec from the Big Bang there was no physics as we understand it, and none of the physical laws we know held. Between 10^{-43} sec and 10^{-34} sec from the Big Bang there were only strong and gravitational interactions, and therefore only their laws held. Between 10^{-34} sec and 10^{-10} sec there were only strong, gravitational, and electroweak interactions, and therefore only their laws held. And so on. The conclusion must be that there is no really universal law, as Reichenbach would like, and as Schlick understood it.

Reichenbach's fourth requirement, concerning unrestricted exhaustiveness, is partially involved in what we have just seen. The basic idea is quite simple. Let us consider a statement, for example $(\forall x)[fx \rightarrow gx]$, and let us put it into what Reichenbach calls the 'D-form', that is, into the disjunctive expansion of its elementary true-cases $(\forall x)[(fx \wedge gx) \vee (\neg fx \wedge gx) \vee (\neg fx \wedge \neg gx)]$. Let us call a 'disjunctive residual' any statement resulting from the original one written in D-form by cancelling one or more terms. At this point, the statement is 'exhaustive in elementary terms' if none of its disjunctive residuals is true.

Since a law of nature must be exhaustive, we can rule out of nomologicity all the statements which have empty antecedents, such as 'all unicorns fly'. In this case the following residual is true: $(\forall x)[(\neg fx \wedge gx) \vee (\neg fx \wedge \neg gx)]$, since it is true that there are no unicorns. It should be noted that, as Salmon (1977, p. 205) pointed out, this requirement appears to be too strong since it also rules out of nomologicity statements such as 'All animals having heart have kidney' since they admit true residuals such as $(\forall x)[(fx \wedge gx) \vee (\neg fx \wedge \neg gx)]$.

However, not only did Reichenbach require nomological statements to be exhaustive, he also required that they were unrestrictedly exhaustive, that is, that there were no space-time regions, for any of their variables, such that a residual was true. In this way he wanted to rule out all those statements which are true and derived from examining all the individuals of a certain kind within the limits of a certain space-time region. Of course this still more demanding requirement is further contradicted by the science we have. For example, we should not consider as laws of nature all the all-statements representing what happens among, and in, the planets of the solar system, since this is limited to a well determined space-time region and to particular kinds of planets possessing given properties. Moreover, any scientific consideration of biological, living beings should be ruled out since all living beings belong to given species, and any species is, or was, made up of a finite number of elements; further, any species occupied or occupies a particular space-time region.

The fifth requirement demands that a nomological statement must be *general in self-contained factors*. Here the matter is a little more complex from a formal point of view. Nevertheless it concerns the requirement that a nomological statement cannot be written as a conjunction of an all-statement and an existential statement. For example, we must eliminate statements such as 'All electrons are charged particles and albino tigers exist', since they can be written such as $(\forall x)[fx \rightarrow gx] \wedge (\exists y)hy$.

At this point, after analysing the formal requirements, we may return to the first (the epistemological requirement): nomological statements must be verifiably true. This is the very core of the matter as far as the status of the laws of nature is concerned: 'being laws of nature, nomological statements, of course must be true; they must be even *verifiably true*, which is a stronger requirement than truth alone' (Reichenbach, 1954, p. 11). That they must be verifiably true is connected with the fact that a statement can be 'factually true', or 'true by chance', as Reichenbach writes, but not a law of nature. The underlying problem concerns, as is evident, the difference between an accidental statement (such as 'all gold cubes are smaller than one cubic mile') and a nomological statement (such as 'all signals are slower than or equally fast as light signals'). In other words, a law of nature should not be conclusively verified, but it must admit possible verification of predictions. Nevertheless, Reichenbach wanted to avoid resorting to modal concepts such as 'possibility', since he feared falling into a circularity.[19] Therefore he preferred to introduce the term 'verifiably true'. This latter has to do with inductive verification interpreted in the sense that a statement is verifiably true if it 'is verified as practically true at some time during the past, present, or future history of mankind' (Ibid., p. 18). Reichenbach pointed out that he did not mean the method of verifying but the set of rules guaranteeing that the inductive verification was used for these statements.

In chapter VI of his 1954 work, Reichenbach tackled a possible objection, according to which what he presented might be good at explicating the notion of 'all existing laws of nature known at some time', since they are highly confirmed at that time, but not of 'all existing laws of nature'. To overcome this objection, he extended the notion of 'verifiably true' to 'verifiably true in the wider sense [...] which makes it permissible to say that there may be laws of nature which will never be found by human beings' (Ibid., p. 85). To reach this aim, he proposed what follows.

He defined:

S_0 = {nomological statements highly confirmed at some time}
P_0 = {observational procedures available at the time at which we define S_0}
R_0 = {observational data resulting from actual or possible applications of P_0}[20]

Then he remarked (Ibid., p. 86):

(1) 'If new laws were discovered then new observational procedures would be constructed';
(2) 'New laws are discovered also by extending the range of known observational procedures'.

According to (1), if we discovered new laws, we could construct new observational procedures. According to (2), if we extended the range of the observational procedures we could discover new laws. Of course 'extending the range of an observational procedure means, logically speaking, using the procedure even when we have no implication stating that a datum of a certain new kind will be observed, but merely know that at least a datum of a known kind will occur, though perhaps trivial' (Ibid., p. 86).

Starting from this second case, he arrived at a definition of a 'possible datum': 'when we speak of possible data, we shall include the use of observational procedures in an extended range' (Ibid., p. 86). But on the previous page he had given a different definition of 'possible datum': 'an occurrence q is physically possible relative to the occurrence denoted by p if neither q, nor $\neg q$ is deductively derivable from p by use the class S_0' (Ibid., p. 85).

As is clear, 'physically possible' in the first sense is totally different from 'physically possible' in the second sense. Anyway, let us go on.

Reichenbach noted that a possible datum is not quantitatively determined by (R_0, P_0, S_0) but it 'depends on the nature of the physical world' (Ibid., p. 86). That is (R_0, P_0, S_0) can permit us to claim that there will be data of a certain kind, but not what their precise value will be. In other words, (R_0, P_0, S_0) 'determines the observational questions we can ask, [but] the answer is given by the physical reality'.

Taking all this into account, we can define:

$$R_1 = \{\text{actual and possible observational data}\}.$$

And therefore, given R_1, we can define:

$$S_1 = \{\text{nomological statements verifiable on the basis of } R_1\}.$$

S_1 is a subset of all the possible laws; in particular it is the subset of the possible laws made up by the statements 'verifiably true on the basis of R_1'. By iterating the above we can arrive at the explication of the notion of 'all possible laws'.

Nevertheless, there are still some problems.

First problem. Does R_1 contain all possible data in the first sense or in the second sense? If we mean the first sense, then R_1 may be considered a good basis on which S_1 can be verifiably true in the wider sense. But if we mean the second sense, then R_1 cannot be considered a good basis on which S_1 can be verifiably true in the wider sense, since it contains both any q and any $\neg q$.

Second problem. If we discovered a law without extending the observational procedures, that is, not in the sense indicated by remark 2, according to remark 1, we would arrive at a set S_1' allowing a set of procedures P_1' which, in turn, permits a set of possible data $R_1' \neq R_1$. But, of course, R_1' would not be a good basis for S_1'.

Third problem. Even if S_1' were discovered by enlarging P_0, we could arrive, as the history of science teaches us, at $R_1' \neq R_1$, where R_1 is obtained via P_0, R_0. It could be the case that there are new data d, derivable from S_1', such that $d \in R_1'$ but $d \notin R_1$, and, therefore, R_1 would not be a good basis for the nomologicity of S_1'.

In conclusion, the notion of 'extension of variability' seems less useful.

To sum up, an all-statement is a law if it is characterized by 'inductive generality', that is, if it is predictive and if such predictions give it a high degree of inductive confirmation.[21] This is nothing but the reformulation of Schlick's criterion of nomologicity. Moreover it has all the problems connected with verification, or, in other words, with confirmation: we cannot know precisely what a law is if we do not know precisely what a confirmation is. This is a very complex point, as the history of contemporary philosophy of sciences teaches us.[22] Now it suffices to emphasize that Reichenbach's strategy started from the logical requirements for a statement to be nomological but that he was then obliged to introduce the methodological requirement of verifiability as well. But, *mutatis mutandis*, this is Schlick's criterion in a new guise. Nothing new, therefore, under the sky of philosophy. Of course not even Reichenbach supported the idea that regularity is intrinsic to nature. It is worth repeating that a law of nature, in this Humean tradition, is not a law that is intrinsic to nature, but a formal description, satisfying certain logical and methodological constraints, of certain regularities observed in nature. Nature is not lawful; instead we must search for the formal lawness.

2.1.3 Nagel: the standardization of Schlick's problem

As we have said, the author who best epitomized the 'standard view' approach to Schlick's problem was undoubtedly Nagel who, in *The Structure of Science*, developed 'an essentially Humean interpretation of the nomic universality' (Nagel, 1961, p. 56). This means that:

> The objective content of the statement that a given event c is the cause of another event e, is simply that c is an instance of a property C, e is an instance of the property E (these properties may be quite

> complex), and C is as a matter of fact also E. On this analysis, the 'necessity' allegedly characterizing the relation of c to e does not reside in the objective relations of the events themselves. The necessity has its locus elsewhere – according to Hume, in certain habits of expectation that have been developed as a consequence of the uniform but *de facto* conjunctions of C and E [...] However, Hume's psychological preconceptions are not essential to this central thesis – namely, that universals of law can be explicated without employing irreducible modal notions like 'physical necessity' or 'physical possibility'.
>
> (Ibid., pp. 55–6)[23]

This was also the epistemological core of Reichenbach's proposal. Unfortunately, this kind of radical anti-metaphysical attitude, in particular, the aversion to solving the question of nomologicity by resorting to modal notions, was not to be followed by many philosophers.

Reading Nagel's great synthesis, we immediately grasp that he was strongly influenced by Reichenbach's result, but with a substantial difference. Nagel was more cautious than Reichenbach, and he did not want to offer a sharp demarcation criterion between nomological conditionals and accidental conditionals. More modestly he wished to furnish some conditions which, more or less plausibly, should be satisfied by any nomological statement.

First of all, a nomological statement is, as is already implicit, a *generalized conditional*. Thus '*it is plausible to require*' (Ibid., p. 59) *that laws are not unrestricted universals,* both in the sense that they should not contain any individual constants and in Maxwell's sense. Nevertheless, as Nagel pointed out, if we really eliminate from nomologicity all the statements containing individual terms we should rule out statements usually regarded as laws. Moreover, he noted that 'different cosmic epochs are characterized by different regularities in nature, so that every statement properly formulating a regularity must contain a designation for some specific temporal period' (Ibid., p. 57). This is exactly the same objection emphasized by Schlick (1931) and Feynman (1965). As far as unrestrictedness is concerned, Nagel very subtly observed that:

> though a universal conditional is unrestricted, its scope of predication may actually be finite. On the other hand, though the scope is finite, the fact that it is finite must not be derivable from the term in the universal conditional which formulates the scope of predication, and must therefore be established on the basis of independent empirical

evidence. For example, though the number of known planets is finite, and though we have some evidence for believing that the number of times the planets revolve around the sun (whether in the past or in distant future) is also finite, these are facts which cannot be deduced from Kepler's first law.

(Ibid., p. 59)

The second requirement concerns the plausibility of eliminating all-statements with empty antecedents, or, as Nagel calls them, those with 'vacuously true antecedents'. Nagel, unlike Reichenbach, recognizes that this is too restrictive, since we have statements considered as laws, but with empty antecedents. For example, 'Copper at −270°C is a good conductor' has an empty antecedent (so long as we are unable to chill copper to −270°C), and the law of inertia has an empty antecedent (we do not have real bodies satisfying it). To maintain such laws with vacuously true antecedents, Nagel suggested that instead of investigating their nomologicity by only considering them as single statements, we should think of them as part of a system of statements. An all-statement with empty antecedent 'counts as a law only if there is a set of other assumed laws from which the universal is logically derivable' (Ibid., p. 60), that is, 'the law is an element in a system of laws for which there are certainly confirmatory instances' (Ibid., p. 62). This is extremely relevant for what I am going to suggest in the third part of this chapter. *To evaluate if an all-statement is a law we must evaluate the entire system of all-statements in which it is inserted.*

The third requirement regards exhaustiveness (note that it is the same term used by Reichenbach): 'it is a plausible requirement that the evidence for it is not known to coincide with its scope of predication and that, moreover, its scope is not known to be closed to any further augmentation' (Ibid., p. 63). Needless to say, again we have Schlick's criterion: a law must not concern only what we know but it must allow for predictions.

The fourth and last requirement again emphasizes the role of the system. For a law, *L*, must have direct and indirect instances. In the former case, the instances fall into the scope of predication of *L*. In the latter case, (a) *L* is jointly derivable with other laws $L_1, \ldots, L_n$ 'from some more general law (or laws) *M*, so that the evidence for these other laws counts as (indirect) evidence for *L*', or (b) '*L* can be combined with a variety of special assumptions to yield other laws each possessing a distinctive scope of predication, so that the evidence for these derivative laws counts as "indirect" evidence for *L*' (Ibid., pp. 64–5). Again, in the indirect case of confirmation, what is relevant is the system in which the all-statement to be considered as a law is included.

To conclude, Nagel indicated that (1) a definitive explication of the notion of law is not possible; nevertheless (2) we must consider some plausible requirements for nomologicity, and (3) we must focus on the system in which the alleged law is situated. This wise and cautious position is an undeniable indication that Nagel was aware of the fact that the 'standard view' approach to laws, uniquely based on the syntactical and methodological analysis of single statements was not sufficient to solve the problem. Which way should be taken, then? Nagel suggests trying the way of counterfactuals: a route on which Goodman had started out only a few years before.

2.1.4 The counterfactual way

The possibility of solving Schlick's problem by resorting to counterfactuality derives from the idea that it seems that only a nomological statement can support a true counterfactual. For example, 'All heated metals expand' is a nomological universal because it supports the truth of the following counterfactual: 'If α were a heated metal, α would expand'. On the other hand, the universal 'All the screws of Jack's car are rusted' is accidental because it cannot support the truth of the corresponding counterfactual 'If β were a screw of Jack's car, then β would be rusted'. In this way the possible solution to the problem of the laws of nature changes. Instead of looking for conditions by means of which to restrain the range of the material implication, attention is focused on the possibility that a universal conditional statement supports a true counterfactual. Doing this means reducing Schlick's problem to the problem of the truth of counterfactuals: a problem that is not at all trivial. It cannot be solved by resorting to classical logic since a counterfactual expressed by a material implication is always true, because its antecedent is false. It cannot be solved by resorting to experience since a counterfactual contradicts facts.

It seems, then, that the condition necessary for a satisfactory analysis of the conditionals implies the overcoming of the paradoxes of material implication, perhaps by constructing an inference between antecedent and consequent independent of the values of the truth of its clauses. Chisholm (1946) moved in this direction, and he thought that the problem at issue could be solved by identifying an additional statement that, added to the antecedent of the counterfactual, would allow us to infer the consequent of the counterfactual. This means introducing that knowledge which would allow us to understand if the counterfactual is true. This was also the way taken by Goodman, first in a formal (and disastrous) way and then in a pragmatic (and successful) way.

With reference to the failure of the formal path, Goodman (1947)[24] came to two conclusions: (a) he showed that the attempt to furnish a good definition of additional conditions involves an endless regression; (b) he pointed out that even though we succeeded in this purpose, we should always establish the kind of connection which would permit the inference of the consequent, and this connection would not be logical, but physical, that is, we should find a law. The result is that the problem of the truth of a counterfactual depends on its being supported by a nomological connection. It means that we can solve the problem of counterfactuals by reducing it to the problem of laws (Schlick's problem). Unfortunately, we saw that the problem of laws should be reduced to the problem of counterfactual. Hence we are in a real vicious circle.

In his 1947 work, Goodman, relating the problem of counterfactuals to Schlick's problem, introduced the issues of induction and confirmation: we have to go back – he writes – to 'the Humean idea that rather than a sentence being used for prediction because it is a law, it is called a law because it is used for prediction; and rather than a law being used for prediction because it describes a causal connection, the meaning of the causal connection is to be interpreted in terms of predictively used laws' (Ibid., p. 21). What characterizes a law is primarily its prediction power, that is, its being valid beyond the observed cases. But *this is, again and mutatis mutandis, Schlick' s criterion*!

So, Goodman goes back to the problem of the laws: if there is a solution, this has to be looked for in the reason why we can move from known to unknown cases. In other words, the problem of induction, together with that of nomologicity (Schlick's problem), is transformed into the 'problem of projection', that is, the problem of defining when a statement is projectable from the observed cases to the not yet unobserved cases.

Thus we arrive at 1953,[25] when Goodman stated that 'by this time we may be ready to try another track' (Goodman, 1954–73, p. 38) in tackling the Humean question of the passage from the known to the unknown, that is, we should face 'the new riddle of induction', and so solve Schlick's problem.

The new riddle of induction concerns the distinction between projectable hypotheses and non-projectable hypotheses. The first aspect emphasized by Goodman concerns exactly what Schlick emphasized. As we recall, Schlick pointed out that any set of empirical data can be expressed by a mathematical representation, and that the same set of data can be expressed by more than one mathematical representation. Goodman begins with the same issue, even if not emphasizing the mathematical formulation but the linguistic representation in general.

In particular, given a set of observed data, he argued that it can be organized by more than one linguistic representation.[26] Among the possible representations, we should consider the relevant ones, that is, the projected representations. Those that, at a certain time, are (1) supported by positive instantiations; (2) inviolate by negative instantiations; and (3) unexhausted. That is, they admit cases which are not negative, but positive and still undetermined. This jargon reformulates Schlick's problem. Which among the possible different projected representations are laws, that is, which are projectable representations? In short, we should demarcate the non-projectable projected representations from the projectable projected representations.[27]

Goodman's solution is pragmatic: 'I think we must consult the record of past projections of the two predicates [representations... and discover which one] as a veteran of earlier and many more projections than [... the other one] has the more impressive biography' (Ibid., p. 94). And thus we are able to understand which is 'much better *entrenched*' than the other one. The notion of 'entrenchment' has a strong pragmatic flavour, since it emphasizes the way in which we speak of the world and the context within which that speaking is performed. This seems a good way out for Schlick's problem. Moreover it is an improvement on the Humean regularism, since it does not focus on the relevance only of past instantiations, but also of how the past instantiations are, and can be, organized by means of linguistic representations. On the other hand, it is difficult not to agree with such a pragmatic position: 'The reason why only the right predicates happen so luckily to have become well entrenched is just that the well entrenched predicates have thereby become the right ones' (Ibid., p. 98). Moreover, it is not a mere question of 'prophecy as a criterion of causality' (Schlick, 1931, p. 184), that is, of law; it is also a problem concerning the status of the representations which allow us to make the 'prophecy'.

Goodman transformed Schlick's criterion into a basis for the pragmatic decision about which regularity must be considered as a law: many regularities are admitted by the same data, among them many are projected; among the projected regularities (which implicitly satisfy Schlick's criterion) only one is actually projectable: the more entrenched. This must be considered as a law. Certainly, this is a solution, but it is pragmatic, and we know that in philosophy the pragmatic solutions are the easiest to discover, but also the easiest to discard, since they are not capable of satisfying a more demanding philosophical taste.

2.1.5 Modality and the new metaphysicians

The failure of classical logic to provide criteria of nomologicity spurred many authors to re-examine the problem of laws in terms of an extremely strong metaphysical commitment based on a rediscovery of modality. On the one hand, the increasing relevance given to the contextual dependence of the truth of the counterfactuals can be thought of as one of the reasons leading towards the modal approach, where the reference to different cognitive contexts is seen in the light of different possible worlds. On the other hand, what characterizes a law is its capability of going beyond the observed cases, its potentiality: a modal property. Thus a number of authors turned their attention to:

(1) The internal relations of universal statements trying either (a) to define the meaning of the modal notion 'necessity' in function of the notion of law (*it is necessary that A, because it is a law that A*); or (b) to define the notion of law starting from the notion of necessity (*it is a law that A, because it is necessary that A*); or
(2) The status of the classes connected with the antecedent A and the consequent C of the nomological statement (*if something is A, then it is necessarily C*), so that they can be interpreted, following the medieval realists, as universals *de re*.

At this point I think it is worth recalling Burks's essay of 1951, 'The Logic of Causal Propositions'. This deserves to be mentioned since it is one of the first attempts to solve Schlick's problem without resorting to classical logic or methodology, but by using modal logic. In particular Burks tried to distinguish the notions of logical necessity $\Box$ from the physical, or causal, necessity $\boxed{c}$, by defining the causal implication, or *c*-implication ($\xrightarrow{c}$), with reference to the usual material implication and to the strict modal implication ($\Rightarrow$). Burks's idea was to bridge these by means of a connection axiom according to which a strict implication materially implies a causal implication, but is not implied by it, that is:

$$(p \Rightarrow q) \rightarrow (p \xrightarrow{c} q)$$

In the same way, we can establish a relation between the causal necessity (possibility) and, on the one hand, the material implication and, on the other hand, the logical necessity (possibility). In the first case, we have

$$\boxed{c}\, p \rightarrow p$$

$$\Box\, p \rightarrow \boxed{c}\, p.$$

Unfortunately, as Burks himself admitted, the *c*-implication carries a causal version of the usual medieval paradoxes of the implication. And not only this. The *c*-implication also involves completely new problems, since it would capture the nomologicity. For example, we can no longer distinguish between two counterfactuals having the same physically or biologically impossible antecedent and opposite consequents. That is, let us suppose we have the following law: 'all massive bodies attract one another' and its violation 'massive bodies over 800 kg do not attract one another'. The latter *c*-implies both 'The Earth and the Moon attract themselves' and 'The Earth and the Moon do not attract themselves'. Obviously, this is a difficult result. It becomes even less pleasing when we consider the important role in science of reasoning from physically or biologically impossible antecedents, as happens in thought experiments (see next chapter). As if this were not enough, Burks's approach does not allow us to deal with the counterfactuals in the contexts in which they are expressed, as other modally-minded philosophers tried to do at that time.

However, in spite of these weaknesses and failures, Burk's work is historically important since it is an attempt to substitute classical logic with modal logic. Unfortunately, as seen, the problems remain, and they are strictly connected with the logical tool.

Now I should move to analyse the weaknesses of the other modal paths mentioned above, were this not totally superfluous to the demolition of any optimism for a metaphysical approach to law and modality which has already been wonderfully made by van Fraassen (1989): the *Alleszermalmender*. I entirely share his *pars destruens* based on the recognition that any metaphysical position on law and modality has at least two serious problems:

> [...] the *problem of inference* and the *problem of identification* [...] The problem of inference is simply this: that it is a law that A, should *imply* that A, on any acceptable account of laws. We noted this under the heading of necessity. One simple solution to this is to equate *It is a law that A* with *It is necessary that A*, and then appeal to the logical

> dictum that necessity implies actuality. But is 'necessary' univocal? And what is the ground of the intended necessity, what is it that makes the proposition a necessary one? To answer these queries one must identify the relevant sort of fact about the world that gives 'laws' its sense: that is the problem of identification. If one refuses to answer these queries – by consistent insistence that necessity is itself a primitive fact – the problem of identification is evaded. But then one cannot rest irenically on the dictum that necessity implies actuality. For 'necessity', now primitive and unexplained, is then a mere label given to certain facts, hence without logical force – Bernice's hair does not grow on anyone's head, whatever be the logic of 'hair'.
> (Ibid., pp. 38-9)

Beyond the fact that if we accept van Fraassen's demolition, there is very little to save in the works of many metaphysicians,[28] their objections to the regularist approach must be borne in mind, even if they underlined little more than the weaknesses that the regularists themselves had already emphasized. In particular the *de re* metaphysicians considered regularism rather 'naive', since it fails to grasp correctly that regularity is neither necessary nor sufficient for nomologicity.[29]

Regularity is not necessary for nomologicity since there are laws that are vacuously true and laws that are spatially and temporarily restricted. According to Armstrong, while the regularists would prefer to restrict nomologicity to the not vacuously true universal statements,[30] the realists still prefer to maintain the vacuously true statements, since the fact that they are not instantiated does not imply that they are not laws. Let us think of a nuclear power station. We know, on the basis of certain physical laws that if certain conditions occur an incident will take place. Therefore, we must act so that these conditions do not occur. In this way, those laws will not be instantiated. Nevertheless this is not a good reason not to consider them as laws. This example also shows that sometimes laws are not absolutely empty, in the sense that because up to now there have not been instantiating cases, that is not to say that this will continue for ever. Sometimes, laws are contingently empty because human intervention prevents their instantiation. In the same way, we cannot a priori deny the existence of local laws, that is, spatially conditioned laws. With reference to this, let us consider Tooley's (1977) bizarre example. Let us think of Smith's garden, where only apples grow. Whenever we try to grow an orange, it transforms into an elephant. Bananas become apples as soon as they enter the garden, while pears cannot pass because of an insuperable force. The cherry trees planted in the garden either produce apples or produce nothing. Should these things really happen, they would be a very strong evidence in favour of the statement 'All the fruits in Smith's garden are apples', even if nothing similar happened in any other garden. To avoid such an evidently local law, we could include it in a more general law. That is, we could affirm that all gardens possessing a certain property have the behaviours mentioned, and that Smith's garden has that property. To confirm this hypothesis we should isolate the property in question, make a garden which has it and confirm whether what is said really happens. But without this confirmation, for Tooley, it would be better to assert the plausibility of a local law, in this case a law valid only for Smith's garden. Though this is a too extreme example for those who prefer philosophical reflection on science rather than on science fiction, it leads to the claim that the presence of a law of nature must not depend on the results of our attempts at significant experiments. These can bring empirical support to the law, but their lack should not induce us simply to deny it.

Regularity is not even sufficient for nomologicity since, notwithstanding how many restrictions we are able to impose, undesired regularities which we cannot consider as laws always remain. The strongest objection against sufficiency, introduced by Kneale (1950 and 1961) and reconsidered by Molnar (1969), concerns the regularists' incapacity to account for empirical non-realized possibilities. We know that a law also tells us something of what has not been observed or not been realized; that is, it should tell us what can and cannot physically happen, independently of its having occurred in a certain space or in a certain time. It is clear that the greatest difficulties arise when we have to do with events of which we do not know any case of instantiation. But it is exactly here that nomologicity should act. Let us consider the two following generalizations: (1) 'All the golden spheres have a diameter inferior to a mile' and (2) 'All the blocks of uranium have a diameter inferior to a mile'. For the regularists, both generalizations are true. But the first is considered as an accidental generalization, while the second is a nomological generalization (there are physical laws which establish the impossibility of a mass of uranium being of such a size). Yet the regularists cannot make such a distinction because they reduce every non-realized possibility to impossibility. For them it is equally impossible that there is a golden sphere with a diameter superior to a mile and that there is a similar sphere of uranium 235: neither situation has ever been realized.[31]

The *pars destruens*, however, is not limited to the extensional difficulties, but considers also the intensional difficulties arising in those cases where a real correspondence between nomologicity and regularity occurs. The difficulties arise from the fact that the regularists do not realize that nomologicity expresses an inner connection between the antecedent and the consequent of the universal statement which is missing in the idea of regularity. As Armstrong (1983, p. 39) pointed out, the simple occurrence of an event which is an *F* and an event which is a *G* does not guarantee the fulfilment of the conditions for nomologicity of 'All the *F* are *G*'. Moreover, even if we had observed a certain number of *F* that are *G*, a regularist could affirm only that 'All the *F* are *G*' means that all the observed *F* are *G*. This is only a description of what happened, not a good reason to affirm that 'All the (past, present, and future) *F* are *G*'.

Something more is needed. But what? Something with a strong metaphysical commitment? Surely we can share the metaphysicians' objections to regularists, even if we already knew them. They show the impossibility of reducing laws to mere regularities, whatever the restrictions imposed. Nevertheless, there is a huge gap between stating this and claiming that the regularities expressed in the law have to be grounded not in external criteria but in some kind of metaphysical necessity either connecting the antecedent and the consequent, or internal to the antecedent and the consequent.

2.1.6 Should we abandon philosophy?

Not only did van Fraassen offer what can be considered the most pregnant and definitive criticism to the metaphysical approach to laws, but he also proposed his own view, based on what, in *The Scientific Image* of 1980, he called 'constructive empiricism', which involves a form of agnosticism substituting the Humean scepticism. From this point of view, we should not look for true theories – either in the realistic sense or in the sense of empiricist reductionism – but for empirically adequate theories. Inside this frame, van Fraassen wanted to 'support the philosophical position that there are not laws of nature' (van Fraassen, 1989, p. 182). A very radical claim! But, of course, much of the supposed radicality depends on what is really meant. In particular he argued that we can have a rationality without laws – as is stated in the title of the second part of *Laws and Symmetry* – in the sense that he proposed a view that science does not need to be interpreted realistically to be considered rational and, therefore, that to believe in the existence of laws is not necessary, in particular to avoid undesired metaphysical flights.

To satisfy this programme, he suggested moving to the notion of 'symmetry': 'I propose that we embark on a study of the structure of science – its theories and models – in itself. The clue, I shall suggest, is this: at the most basic level of the theorizing, *sive* the model construction, lies the pursuit of symmetry' (Ibid., p. 233).

Regularity in nature, according to van Fraassen, is not grasped by means of laws, whose ontological weight leads erroneously to postulating the discovery of realities beyond the observable, but by means of concepts like 'symmetry' and 'covariance'. It is this idea, already presented along nearly the same lines in 1962 by Pap in *An Introduction to the Philosophy of Science*, which allowed van Fraassen to develop his 'constructive' version of empiricism. Van Fraassen denied the possibility that there is a correspondence between the structure of scientific formulations and the structure of the world, which, as we have seen, is an assumption of many interpretations of the laws of nature. Rather, he focused on the former – scientific formulations – especially where they were 'empirically adequate'. To realize this idea he identified as central the concept of 'symmetry'. The concept of symmetry does not remain a vague intuition, but is formally specified. Symmetries concern a set of transformations leaving invariant a certain structure, even though the context within which it is considered changes. We can show that the set of these transformations makes up an algebraic group. This is important, since it shows how the attempt to catch the covariance through symmetry is not only intuitively stimulating but it has a formal counterpart in the mathematics field, where it demands a precise algebraic structure.

In synthesis, we could say that to explain the regularities we must resort to covariance. In other words, generalizations depend on the structure of the formalizations, for 'the equation is *covariant* exactly if it is either true for all the frames of reference or for none' (Ibid., p. 281). It is this notion of 'covariance' that should permit us to explicate the vague notion of 'generalization', which we require to characterize any statement supposed to be a law. However, does the fact of discussing formal structures instead of nature or world and of covariance instead of generalization lead us to a definite solution to the problem of laws of nature? Van Fraassen's answer is as honest as it is disconcerting: 'No!'. Claiming that a statement is covariant is too little, as we can understand by reflecting on the fact that the statement 'The planets of the solar system are nine' is covariant but absolutely inconceivable as a law (Ibid., p. 288).

Perhaps we have found a good explication of the notion of 'generalization' (at least in physics) if not of lawness. But it seems much ado about nothing. Certainly van Fraassen's philosophical criticism is very powerful, but his proposal is a sort of withdrawal from philosophy; every conception of law that has been advanced, especially if it has been advanced by metaphysicians, has to be considered as non-valid; the only thing we can do is to abandon the attempt to grasp the lawness philosophically and to search for a possible solution beyond philosophy, in particular in the structure of scientific formalizations. Between the Scylla of the Humean regularism and the Charybdis of the pre-Humean metaphysics, I think we have found an abandonment of the philosophical task.

If we reflect on the debate concerning the laws of nature, we realize that the neopositivist movement shifted the focus from the role and the cognitive meaning of laws to the problem of the logico-epistemological conditions characterizing a universal statement as a nomological statement. This, that is, Schlick's problem, has been at the heart of the question of the laws for the neopositivists and for the post-positivists. However, in the end, the debate on the conditions has transmuted into something else. It has become an ontological debate: with on the one hand, the new metaphysicians believing in the ontological anchorage of the regularities indicated by the laws, and, on the other hand, the new regularists who, being the direct descendants of the neopositivist aversion for any ontological or metaphysical issue, deny that anchorage. Nevertheless, both the old and the new regularists failed to give a good account of their position, and the new metaphysicians failed as well. Why, then, not turn to the 'stone guest', that is, to Kant, and look for a breath of fresh air in his tradition?

2.2 Back to Kant

2.2.1 The *status quaestionis*

Dealing with the topic of laws of nature in Kant's critical works is not an easy task: there are numerous difficulties within Kant's writings, and many problems related to the secondary literature. As far as Kant's writings are concerned, the difficulties are particularly in regard to the connection between the various levels within which discussion of laws is possible. There are: (1) the transcendental level, almost exclusively tackled in the *Critique of Pure Reason* (1781–87); (2) the metaphysical level, analysed in the *Metaphysical Foundations of Natural Science* (1786); and (3) the empirical level, particularly discussed in the *Critique of Judgement* (1790). As far as the most influential secondary sources are concerned, they present an astonishing division: on the one hand, there is the major exegetic literature before 1960, which is in German, French, English and Italian; on the other hand, there is the major exegetic literature after 1960, which is almost entirely in English. In fact the first group does not deal with the problem of laws of nature as an independent topic; rather this is embedded in wider discussions about the transcendental deduction, the analytic of principles, the second analogy, the metaphysical principles of physics and the teleological judgements. Moreover, it is a literature which, apart from Kant's writings, covers about two centuries of historical tradition, thus including the most important representatives of Kantian scholarship, such as Cohen, Cassirer, Vaihinger, Adickes, Kemp Smith, Paton, de Vleeschauwer and Scaravelli. The second group focuses specifically on the problem of laws, which is analysed more rigorously from the logical point of view. It is a literature which sometimes concerns only a given volume of Kant's writing, sometimes only a given part of a given volume. Further, it is a literature which particularly takes into account the Anglo-Saxon debate of the last twenty or thirty years. This means, first, that the great (continental) European Kantian tradition is sometimes simply ignored, and second, that those dealing with the problem of laws can seem to belong to a sort of secret society where each member especially mentions another.

Undoubtedly the first literature has its drawbacks, particularly those regarding the absence of any direct focus on the problem of the laws of nature. But, although meritorious for example in individuating the problem and in using appropriate interpretative tools deriving from logic and the philosophy of science, the second literature also has some flaws. There are those who interpret Kant in a too strongly emphasized contemporary fashion; those reading him with Hume's eyes; those missing the architectonics of his philosophical system; those interpreting him as an inductivist.[32] I will try to mediate between the two positions, that is, on the one hand, I will take into account a wide exegetic context, and, on the other hand, I will analytically focus on the problem of law.

Nevertheless, before beginning my analysis, I would like to specify some lexical aspects, especially related to the terms *Gesetzlichkeit* and *Gesetzmässigkeit*. Both come from *Gesetz*, a term which in eighteenth-century German and, thus, also in Kant's language, means 'law' in the sense of something imposed. Thus, it differs from *Recht*, that is, from 'law' seen as a command resulting from an agreement between independent individuals. In other words, the difference between *Gesetz* and *Recht* is precisely the Latin difference between *lex* and *jus*. So, not accidentally, Kant uses *Gesetz* when talking about the law of nature and the moral law, and *Recht* when dealing with political issues (see Krieger, 1965).

Both *Gesetzlichkeit* and *Gesetzmässigkeit* are used by Kant in the sense of 'to be in accordance with a law', 'to be allowed by a law'. Nevertheless, since the first is present only six times in all his writings,[33] this distinction is not particularly useful. In fact, there are two different questions which Kant, because of obvious historical contextual limitations, does not distinguish. The first one concerns the conformity to a law of something extralinguistical, that is, its *lawfulness*. In this sense, nature is lawful, it is in accordance with a law. The second regards a different aspect and it is related to the universal statements. As we have seen, while each law is a generalization, not each generalization is a law; it is a law only if it satisfies certain requirements, and therefore we can speak of its *lawness*.

Both these problems are present in Kant, but the first, concerning lawfulness of nature, is explicitly discussed, while the other, regarding lawness, that is, the nomological validity, of lawlike statements, is discussed only implicitly. Moreover, as I will show, the issue of the lawfulness is present at the transcendental, the metaphysical and the empirical level. At each level, of course in different ways, the question of whether, and in what extension, nature is lawful is analysed. The issue of nomological validity, however, is present only at the empirical level, and it could not be otherwise. It suffices to recall that it is related to the distinction between accidental statements and nomological statements. This is not a problem at the transcendental level, because at this level which statements are to be considered laws is well known: those which make possible an application of the categories, that is, the pure principles of the understanding. Nor is it a problem even at the metaphysical level: here the statements considered to be laws are precisely the metaphysical principles. The question takes up a different aspect at the empirical level, where it is not so trivial, differentiating the empirical laws, that is, the nomological statements, from the accidental statements.[34]

2.2.2 The transcendental level

The lawgiving understanding

Let us start by recalling that Kant considers the understanding (*Verstand*) from many different angles: as spontaneity of knowledge, faculty of thinking, faculty of concepts, faculty of judging, faculty of knowledge and faculty of rules. For the understanding has the faculty of producing the representations from itself. Moreover, if it is considered as faculty of thinking, since thinking, in this sense, means knowing by concepts, it is also both faculty of knowing and faculty of judging, because concepts are predicates of possible judgements. (This is the *Leitfaden* that makes it possible to infer the table of categories from the table of judgements). Finally, it is faculty of rules, as concepts are rules.

Note that, in Kant's view, neither the faculty of judging, nor the faculty of rules are equivalent to the faculty of subsuming under a certain judgement, or under a certain rule: this is the faculty of judgement. For the faculty of judgement (specifically of determining judgement, as we will see) concerns *casus datae legis*.

Reason (*Vernunft*) is also a faculty; more precisely the 'faculty of principles' (KdrV, p. 301, B356). Here we should be alert. As is known, the source of the principles, at least the pure ones, is the understanding (Ibid., pp. 194–5, B197–8). However, what exactly does Kant believe the faculty of principles to be? This question can be interpreted in two ways, which, however, converge. First, since 'understanding may be regarded as a faculty which secures the unity of appearances by means of rules, [...] reason [... may be seen as] the faculty which secures the unity of the rules of understanding under principles' (Ibid., p. 303, B359). And, of course, these are not the principles of the understanding, but those belonging to reason. To grasp the second one, a distinction between the two meanings of the term 'principle' must be recalled: (1) 'principle' as general knowledge functioning as a premise of a mediate inference (a syllogism), that is, knowledge from which the deduction begins; (2) 'principle' as unconditioned, as condition without being conditioned (Ibid., pp. 301–2, B356–8). Now, it is precisely on the basis of this unconditioned that is provided 'to the manifold knowledge of the latter [the understanding] an *a priori* unity by means of concepts, a unity which may be called the unity of reason, and which is quite different in kind from any unity that can be accomplished by the understanding' (Ibid., p. 303, B359).

In order to understand which the concepts and the principles of reason are, we should begin with the logical use of reason, that is, from what makes possible a passage from statements to statements through the syllogistic inferences. Here two aspects must be considered: the *direction* of the syllogistic inference and its *form*. As far as the direction is concerned, we can 'reason' either in a epysillogistic way, that is, extracting the consequences one by one, and, thus, working on the side of the conditioned; or we can 'reason' in a prosyllogistic way, that is, on the side of the conditions. In the first case, there are no problems and particular knowledge is obtained; in the second case there are evident problems: either an unconditioned condition is found, or we have an infinite regression:

> but however this may be, and even admitting that we can never succeed in comprehending a totality of conditions, the series [of premises] must none the less contain such a totality, and the entire series must be unconditionally true if the conditioned, which is regarded as a consequence resulting from it, is to be counted as true. This is a requirement of reason, which announces its knowledge as being determined *a priori* and as necessary, either in itself, in which case it needs no grounds, or, if it be derivative, as a member of a series of grounds, which itself, as a series, is unconditionally true.
>
> (Ibid., p. 322, B389)

Therefore, it is a need of reason, or a *requirement of reason* (*Forderung der Vernunft*), given by its logical use, to find out an unconditioned condition. And this need can be formulated in a principle: 'to find for the conditioned knowledge obtained through the understanding the unconditioned whereby its unity is brought to completion' (Ibid., p. 306, B364). In brief: 'Hence, just as the understanding needs the *categories* for experience, reason contains in itself the basis for *ideas*, by which I mean necessary concepts whose object nevertheless *cannot* be given in any experience' (P, p. 82, 328).

However, Kant does not discover any other principle of reason on the grounds of this 'principle peculiar to reason in general, in its logical employment' (Ibid.). Nevertheless, as the categories are 'deduced'[35] from the logical forms of judgement, so the concepts of reason, that is, the ideas (and, therefore, the principles of reason), are now 'deduced' from the logical form of reasoning, that is, from the forms of syllogism: categorical syllogism, hypothetical syllogism, and disjunctive syllogism. In this way, respectively, he obtains the transcendental ideas of (1) the absolute subject: the 'soul' (the '*unconditioned* [...] of the *categorical* synthesis in a *subject*'); (2) the absolute object: the 'world' (the unconditioned 'of the *hypothetical* synthesis of members of a *series*'); (3) the absolute ideal: 'God' (the unconditioned 'of the *disjunctive* synthesis of parts of a *system*', KdrV, p. 316, B379).

As is known, an erroneous use of pure reason implies a dialectic, that is, the illusion that those ideas have an objective validity and not a mere subjective validity.[36]

Having traced the difference between understanding and reason, I want to focus on an aspect of understanding particularly relevant to what I am discussing: its being a legislator. At this point it is worth recalling that in the mid-1700s, philosophers addressing the epistemological problem of knowledge-formation inquired substantially in two possible ways:

(1) the object is given to us inductively, in order to be represented in its theoretical bareness;
(2) the knowing subject knows the object producing it completely through its own representations (see Ibid., pp. 174–5, B167–8).

Kant, as we read in his letter of 21 February 1772 to Marcus Herz, rejects both the first possibility, which would imply an *intellectus ectypus* (a sort of derivative intellect), and the second, which would imply an *intellectus archetypus* (a sort of original intellect), which is a purely divine feature.[37] Instead, he proposes a third way between the radical kind of empiricism (the first possibility just mentioned), and extreme rationalism (the second possibility). This solution involves there being 'two stems of the human knowledge' (Ibid., p. 61, B29): sensibility and understanding. In this way, 'experience [. . .is] a species of knowledge which involves understanding; and understanding has rules which I must presuppose as being in me prior to objects being given to me, and therefore as being *a priori*. They find expression in *a priori* concepts to which all objects of experience necessarily conform, and with which they must agree' (Ibid., pp. 22–3, BXVII). Now we are at the very heart of Kant's *Copernican revolution*,[38] according to which:

> reason has insight only into that which it produces after a plan of its own, and that it must not allow itself to be kept, as it were, in nature's leading-strings, but must itself show the way with principles of judgment based upon fixed law, constraining nature to give answer to questions of reason's own determining. Accidental observations, made in obedience to no previously thought-out plan, can never be made to yield a necessary law, which alone reason is to discover. Reason, holding in one hand its principles, according to which alone concordant appearances can be admitted as equivalent to laws, and in the other hand the experiment which it has devised in conformity with these principles, must approach nature in order to be thought by it. It must not, however, do so in character of a pupil, who listens to everything that the teacher chooses to say, but of an appointed judge who compels the witnesses to answer questions which he himself has formulated.
>
> (Ibid., p. 20, BXIII)

All the above leads to the idea that human understanding imposes its laws a priori – its pure principles – on the world, constituting it as nature in its lawful character:

> that the highest legislation for nature must lie in ourselves, that is, in our understanding, and that we must not seek the universal laws of nature from nature by means of experience, but conversely, that we must seek nature, as regards its universal conformity to law, solely in the conditions of the possibility of experience that lie in our sensibility and understanding.
>
> (P, p. 73, 319)

> *the understanding does not draw its (a priori) laws from nature, but prescribes them to it.*
>
> (Ibid., pp. 73–4, 320)

In brief, the human understanding coordinates, but primarily synthesizes as far as knowledge production is concerned, the representations of empirical intuition, ordering them, and producing in this way the lawfulness of experience. This legislating faculty, or capacity, is conferred on it by its own 'true primary concepts' (KdrV, p. 114, B107), that is, the pure concepts or, as Kant, resorting to Aristotle, calls them, categories (Ibid.). Certainly, these must be applied, but this is possible through the pure principles of the understanding (*die Grundsätze der reinen Verstand*), which are in this way the rules of subjective use. Therefore applying a pure concept, for example, that of cause, to a phenomenon means subsuming the latter under the corresponding rule (the pure principle), in our case that for which any cause is followed from an effect. However, to apply a pure concept, that is, to subsume a phenomenon under a principle, requires that the gap between understanding and sensibility is filled. And, as we know, this is made thanks to the different schemata related to the corresponding categories.

Kant points out that 'principles *a priori* are so named not merely because they contain in themselves the grounds of other judgments, but also because they are not themselves grounded in higher and more universal modes of knowledge. But this characteristic does not remove them beyond the sphere of proof' (Ibid., p. 188, B188). Note that their justification cannot be found by resorting to something superior to them, exactly because they are the superior in question. However, a justification (*Beweis*) cannot be avoided. The way followed by Kant is to prove their necessity on the basis of 'a proof, from the subjective sources of the possibility of knowledge of an *object in general*' (Ibid.). That is, he argues that without those principles, we could not have any knowledge of objects.[39]

As the classes of the categories are four, so the classes of the principles are four. Moreover, as the schemata are eight, one for each of the first two classes of categories and three for each of the other two classes, so there are eight corresponding principles.

The first two classes of principles are the *constitutive principles*, and the other two classes are the *regulative principles*. Between the latter, the *analogies of experience* are particularly interesting for us. They have the following basic principle: 'experience is possible only through the representation of a necessary connection of perceptions' (Ibid., p. 208, B218). Here we are facing laws, since we are dealing with 'necessary connection'. For the phenomena can be related to each other in many arbitrary and subjective ways described by the judgements a posteriori. But thanks only to the imposition of the categories of relation, these associations can receive the necessity and universality characterizing the laws. How this happens can be synthesized as follows:

> the relation [involved] in the existence of the manifold has to be represented in experience, not as it comes to be constructed in time but as it exists objectively in time. Since time, however, cannot itself be perceived, the determination of the existence of objects in time can take place only through concepts that connect them *a priori*.
>
> (Ibid., p. 209, B218)

Since the modes of time are three (duration, succession and coexistence), as we know from the three schemata (permanence, succession and simultaneity) making the application of the categories of relation possible, we have three synthetic principles a priori, which specify, regarding the three schemata and the corresponding three categories, the basic principle before mentioned.

I do not dwell upon the first analogy of experience (that of the *permanence of the substance*), whose basic principle claims: 'in all change of appearances substance is permanent; its quantum in nature is neither increased nor diminished' (Ibid., p. 212, B224). While I postpone for a while a detailed analysis of the second analogy, for which '*All alterations take place in conformity with the law of the connection of cause and effect*' (Ibid., p. 218, B222), I wish to briefly consider the third analogy, whose basic principle states that '*All substances, in so far as they can be perceived to coexist in space, are in thoroughgoing reciprocity*' (Ibid., p. 233, B256).

This principle completes the legislation of nature: if the second gives a rule for the interactions which do not occur at the same time, this provides a rule for the interactions which occur at the same time, that is for the coexistence of the phenomena.

In apprehending multiplicity, we are often dealing with phenomena whose perception is temporally indifferent: we can perceive first the phenomenon *A* and then the phenomenon *B*, or the other way round. In that case *A* and *B* are contemporaneous. The mere apprehension would lead only to an arbitrary coexistence of *A* and *B*. But, whether the pure concept of 'community' intervenes, that subjective association turns into an objective one, that is, it becomes universal and necessary. This means that there is a transcendental condition a priori, specified in the principle of the third analogy, by means of which to have knowledge of what coexists is possible. Of course, this means that coexistence should not be thought in the phenomena, but in the laws a priori of the understanding that make the relation among them (the phenomena) cognitively significant.

It is worth mentioning that the last two principles speak about the relations among the phenomena, while the first analogy makes such relations possible. For we can talk about succession (in the second analogy) and about coexistence (in the third analogy) only because there is a substance which remains, and with reference to which we are able to differentiate the two temporal determinations. In a certain sense, the first analogy can be thought of as a necessary condition for the other two.

Nature überhaupt *and experience* überhaupt

As we have said, understanding imposes the categorial apparatus, that is, its pure principles, to nature so as to cognitively constitute it as an object of our knowledge. This implies two consequences: (1) knowledge of the phenomena (or of relations among phenomena) is knowledge of representations (or of relations among representations, that is, representations of representations); (2) experience is not possible without the intervention of the understanding: the source of spontaneity of knowledge. The cognitively constitutive imposition of laws to nature means that what is known is known in the light of a certain law. In other words:

(1) in nature 'nothing happens through blind chance (*in mundo non datur casus*)' (Ibid., p. 248, B280), but all happens because there

is a certain rule a priori (this is a consequence of the principle of causality);

(2) 'no necessity in nature is blind, but always a conditioned and therefore intelligible necessity (*non datur fatum*)' (Ibid.) (this is a consequence of the fact that not only is there causality, but also that it is necessary);

(3) in nature there are no gaps: '*in mundo non datur saltus*' (Ibid., p. 248, B281) (this is a consequence of the anticipations of perceptions);

(4) in nature there is no discontinuity: '*non datur hiatus*' (Ibid., p. 249, B281) (this is still a consequence of the anticipations of perceptions).

Note that the Transcendental Analytic, the Analytic of Principles included, may be seen as a long argument for the claim that *the understanding a priori anticipates the form of possible experience 'in general'* (*überhaupt*). This 'form' of possible experience 'in general' is nothing but the *regularity (lawfulness) of the phenomena in space and time* (see Scaravelli, 1968, p. 292). Therefore the lawfulness of *possible nature 'in general'* (*überhaupt*) is the result of the legislating understanding.

At this point, it should be easy to grasp the true meaning of the principle of the synthetic a priori judgements, according to which 'every object stands under the necessary conditions of synthetic unity of the manifold of intuition in a possible experience' (KdrV, p. 194, B197), that is, 'the conditions of the *possibility of experience in general* are likewise conditions of the *possibility of the* objects *of experience*' (Ibid.). For the conditions of the possibility of experience are exactly the pure principles of the understanding, that is, what makes both the cognitive constitution of objects and the cognitive constitution of the relations among them possible.

It follows immediately that nature is not something rough given to man, but something completely cognitively constituted both (1) regarding the elements of which it is composed, that is, phenomena (that are representations), and (2) regarding the nomological relations connecting those phenomena (that are still representations). In brief, it is to be thought of both (1) as *natura materialiter spectata*, that is, as the class of the phenomena *constituted* as such by the forms of space and time and then by the categories, and (2) as *natura formaliter spectata*, that is, as the class of the nomological relations among phenomena (see Ibid, pp. 172–3, B163–5) *constituted* by the pure principles of the understanding, especially by the analogies of experience.[40] It follows that the transcendental question, 'How is nature "in general" possible?', must be divided in two different questions:

> *First*: How is nature possible in general in the *material* sense, namely, according to intuition, as the sum total of the appearances; how are space, time, and that which fills them both, the object of sensation, possible in general? The answer is: by means of the constitution of our sensibility, in accordance with which our sensibility is affected in its characteristic way by objects that are in themselves unknown to it and that are wholly distinct from said appearances [...] *Second*: How is nature possible in the *formal* sense, as the sum total of the rules to which all appearances must be subject if they are to be thought as connected in an experience? The answer cannot come out otherwise than: it is possible only by means of the constitution of our understanding, in accordance with which all these representations of sensibility are necessarily referred to a consciousness, and through which, first, the characteristic mode of our thinking, namely by means of rules, is possible, and then by means of these rules experience is possible – which is to be wholly distinguished from insight into objects in themselves.
>
> (P, pp. 72–3, 318)

Let us focus on the meaning of 'in general' (*überhaupt*), since it is this qualification that shows the particular features of the transcendental level. We know that 'transcendental' concerns 'all knowledge which is occupied not so much with objects as with the mode of our knowledge of objects in so far as this mode of knowledge is to be possible *a priori*' (Ibid., p. 59, B25). Dealing with the possibility of nature 'in general' means dealing with the way according to which it is made possible a priori. On the one hand, through the categories of quantity and quality, and, thus, on the grounds of the pure principles of the understanding that are called 'constitutive', the knowing subject constitutes the possibility of the elements of nature – the phenomena – making, in this way, knowledge of them possible, that is, making them possible both (1) as extensive quantities in space and time (by the axioms of intuition), and (2) as intensive magnitudes possessing a degree of perceptible reality (the anticipations of experience). On the other hand, especially through the categories of relation, and, thus, on the basis of the corresponding pure principles of the understanding that are called 'regulative', the knowing subject imposes the main relations among those elements, that is, imposes that (1) in the various relations among the phenomena the total 'substance' does not change; (2) there is a rule (of a causal type) for their diachronic relations; (3) there is a rule (of a reciprocal type) for their synchronic interactions.

A last step regards the determination of the relation between possible experience 'in general' and possible nature 'in general'. Here, it is worth mentioning that, for Kant, there are at least two meanings of the term 'experience': (1) one regarding 'with'[41] what knowledge begins; (2) one concerning the realization of knowledge through the matching between the empirical datum and the categorial apparatus.[42] It is this second meaning which is now interesting for us:

> The possibility of experience in general is therefore at the same time the universal law of nature, and the principles of the former are themselves the laws of the latter. For we are not acquainted with nature except as the sum total of appearances, that is, of representations in us, and we cannot therefore get the laws of their connection from anywhere else except the principles of their connection in us, that is, from the conditions of necessary unification in one consciousness, which unification constitutes the possibility of experience.
>
> (P, p. 72, 319)

Taking into account what was said on nature 'in general' and on the legislating understanding, it is possible to conclude that *natura formaliter spectata*, since it is the product of the imposition of the pure laws of the understanding, especially of the analogies of experience, it is nothing but a different way of considering the lawfulness of nature 'in general', that is, the lawfulness of nature at the transcendental level.

2.2.3 The metaphysical level

The problem of the metaphysical principles

While we have considered the pure principles of the understanding (*die Grundsätze der reinen Verstand*) at the transcendental level, at the metaphysical level the first metaphysical principles (*die metaphysische Anfangsgründe*) must be faced. Immediately we meet one of the main problems raised by contemporary scholars: are the first principles really deduced from the pure principles? Is there any strict logical connection between the transcendental level and the metaphysical level? Some commentators claim that the first metaphysical principles are neither logically obtained from the pure principles, nor justified by the table of categories, which, on the contrary, supplies only a classifying guide (see in particular Buchdahl, 1965; defended by Allison, 1994). But others claim that there is a strict logical and epistemological relation between the two classes of principles (see Friedman, 1994;

defended implicitly by O'Shea, 1997). Finally, there are commentators who, although pointing out some inconsistency between the *Critique of Pure Reason* and the *Metaphysical Foundations of Natural Science*, are inclined to neglect the metaphysical in favour of the transcendental Kant (see Lee, 1981, in particular, pp. 400–4). I must enter this debate, but, first, I want to recall some traits of the metaphysical approach here in question.

Kant, both in Chapter III of the Doctrine of Method and in the Introduction to the *Metaphysical Foundations* gives a taxonomy of the different philosophical fields. First he divides *pure philosophy*, which is 'knowledge obtained by reason from empirical principles' (KdrV, p. 659, B868) and *empirical philosophy*, deriving from the empirical concepts. Pure philosophy can be intended either (1) in the propaedeutic sense of '(preparation), which investigates the faculty of reason in respect of all its pure *a priori* knowledge' (Ibid., p. 659, B869); it is called then *critical philosophy* or *transcendental philosophy*; or (2) as 'the system of pure reason, that is, the science which exhibits in systematic connection the whole body (true as well as illusory) of philosophical knowledge arising out of pure reason' (Ibid.); this is called *metaphysics*. According to this division, I have been dealing with pure philosophy in the propaedeutic sense, that is, with transcendental philosophy. Now, I must move to that part of pure philosophy considered as the science of the pure principles, that is, metaphysics. However, metaphysics is also divided into two parts: that regarding the speculative employment of pure reason, which is the *metaphysics of nature*,[43] and that concerning the practical employment of pure reason, which is the *metaphysics of morals*. Finally, the metaphysics of nature, called also 'physiology' (Ibid., p. 662, B873), is divided further into the 'doctrine of bodies' – as Kant calls it in the *Metaphysical Foundations* – which deals with the objects of external senses and the 'doctrine of the soul' – as it is called in the same work – which deals with the objects of the internal sense. Now, 'the metaphysics of corporeal is entitled *physics*' (Ibid., p. 662, B874) or '*physica rationalis*' or '*physicam puram*' (Ibid., p. 56, fn. A, B21).

Note that Kant gives a special position to physics, in the sense that, unlike chemistry, biology and psychology, it has a pure part which is written in mathematical language. This is extremely important since the more mathematics a discipline contains, the more it is a '*proper* science' (M, p. 6, 470); where (1) by 'science' Kant intends knowledge ordered in a system, (2) 'proper' stands for the fact that its principles have an apodictic certainty (Ibid.), that is, they are synthetic a priori judgements.[44] With reference to the ordered system, notice that if it is given in what we would call an almost axiomatic way, that is, if it is logically structured with principles and consequences, then we have a *rational science* (Ibid., p. 4, 468).

A problem arises from the above. Since both the principles of pure understanding and the metaphysical principles are synthetic a priori judgements, what is the difference between them? In order to answer this question, it is necessary, but not sufficient, to grasp the difference between *general metaphysics* dealing with nature 'in general', and the *particular metaphysics* concerning a nature in particular. In our case (we are considering only natural sciences) this is composed of the objects of the external senses:

> *Properly* so-called natural science presupposes, in the first place, metaphysics of nature. For laws, that is, principles of the necessity of that which belongs to the *existence* of a thing, are concerned with a concept that cannot be constructed, since existence cannot be presented *a priori* in intuition. Thus proper natural science presupposes metaphysics of nature. Now this latter must always contain solely principles that are not empirical (for precisely this reason it bears the name of metaphysics), but it can still either; *first*, treat the laws that make possible the concept of a nature in general, even without relation to any determinate object of experience, and thus undetermined with respect to the nature of this or that thing in a sensible world, in which case it is the *transcendental* part of the metaphysics of nature; or *second*, concern itself with a particular nature of this or that kind of thing, for which an empirical concept is given, but still in such a manner that, outside of what lies in this concept, no other empirical principle is used for its cognition [...], and here such a science must still always be called a metaphysics of nature, namely, of corporeal [...] nature [...], but a *special* metaphysical natural science (physics [...]), in which the above transcendental principles are applied to the [...] objects of our [external] senses.
>
> (Ibid., pp. 5–6, 469–70)

Therefore, while general metaphysics deals with nature 'in general', the metaphysics of one given science deals with a nature in particular, that is, with that particularization of nature limited to the particular class of objects considered by that given science. However, this does not mean that general metaphysics and particular metaphysics are independent. In fact, the principles of the first, related to the possibility of an object 'in general', make it possible for this object to be investigable by physics, in particular by pure physics.

The transition from the transcendental to the metaphysical level is enabled by the introduction of a specific concept which allows us to delimit the research in question to a specific field: either that of physics, or that of psychology. In the case of physics, since we are dealing with objects affecting the external senses, the empirical concept to be introduced is that of 'matter' (see also KdU, pp. 20–1, 181–2 – II Introduction). This 'matter' is not any particular matter, but 'matter in general (*überhaupt*)' (M, p. 11, 475).

Now we can conclude that passing from the pure principles to the metaphysical principles, we do not have a change in epistemological status, since both are synthetic a priori judgements, but that there is a change in the epistemological rank. Although both classes are formed by universal statements, they have a different 'precedence in respect of generality' (KdrV, p. 660, B871): the pure principles have for their object nature 'in general', while the metaphysical principles have a particular physical nature.

The mathematical lawfulness of physical nature

We have seen that it is necessary, but not sufficient, to grasp the difference between pure principles and metaphysical principles in order to specify the difference between general metaphysics and particular metaphysics. For this difference allows us to understand the distinct rank of the two classes of principles, but it is not sufficient to grasp their epistemologically different roles, and, thus, to characterize the transcendentality of the first and the metaphysicity of the second. We already know what the transcendentality of the first is, but what is the metaphysicity of the second?

To answer we should take into consideration mathematics, that is, what characterizes rational physics, that is, pure physics, the metaphysics of nature.[45] As we know, the transcendental level deals with the object 'in general', or rather, with the possibility of the object 'in general', while the metaphysical level deals with a particular type of object, in our case, those of the external senses. However,

> the possibility of determinate natural things cannot be cognized from their mere concepts; for from these the possibility of the thought (that it does not contradict itself) can certainly be cognized, but not the possibility of the object as a natural thing that can be given outside the thought (as existing). Hence, in order to cognize the possibility

> of determinate natural things, and thus to cognize them *a priori*, it is still required that the *intuition* corresponding to the concept be given *a priori*, that is, that the concept be constructed. Now rational cognition through construction of concepts is mathematical.
>
> (M, p. 6, 470)

Often, but especially in the Transcendental Doctrine of Method, in the section entitled 'The Discipline of the Pure Reason in Its Dogmatic Use', Kant claims that while philosophical knowledge is discursive, that is, based on concepts, mathematical knowledge is based on the construction of concepts, where:

> to construct a concept means to exhibit *a priori* the intuition which corresponds to the concept. For the construction of a concept we therefore need a *non-empirical* intuition. The latter must, as intuition, be a *single* object, and yet nonetheless, as the construction of a concept (a universal representation), it must in its representation express universal validity for all possible intuitions which fall under the same concept.
>
> (KdrV, p. 577, B741)

That is why mathematics is so successful in physics. Physics also concerns objects without having an empirical intuition of them. This can be positively managed only by constructing their concepts, which is possible only by using mathematics. If one reflects on this aspect, one become aware that Kant is affirming something which many of us can share: doing theoretical physics means representing objects conceptually, that is, without taking into account the empirical aspects. This is possible exclusively on the grounds of mathematics. Having conceded that mathematics allows us to construct the concepts of physical objects without having an intuition of them, we should understand how this happens. Kant has a suggestion:

> But in order to make possible the application of mathematics to the doctrine of body, which only through this can become natural science, principles for the construction of the concepts that belong to the possibility of matter in general must be introduced first.
>
> (M, p. 8, 472)

These 'principles of the *construction* of the concepts' are the first metaphysical principles of natural science. It means that by analysing the concept of matter 'in general' those principles a priori allowing us the application of mathematics and, thus, the construction of the concepts of rational physics, can be found. This is the reason why 'all natural philosophers who have wished to proceed mathematically in their occupation have always, and must have always, made use of metaphysical principles (albeit unconsciously), even if they themselves solemnly guarded against all claims of metaphysics upon their science [...] Thus these mathematical physicists could no way avoid metaphysical principles' (Ibid.).

Therefore, the 'possibility of a mathematical doctrine of nature' is based on the 'principles of the construction of these concepts' (Ibid., p. 9, 473) of objects of the external senses, independently of their actual intuition; that is, as it is written in the *Critique of Judgement*: 'a principle is called metaphysical if it is one [by] which [we] think the *priori* condition [thanks to mathematics, we could add] under which alone objects whose concept must be given empirically can be further determined *a priori*' (KdU, pp. 20–1, 181).

If the reading I have just proposed is valid, then Buchdahl is right in saying that there is no strict deduction of the metaphysical principles from the pure principles. (There is no mention of any type of deduction in Kant's 1786 work.) In fact there is a connection between the two classes of principles: without the pure principles it would not be possible to have knowledge of objects whose concepts are mathematically constructed by physics. There is also another connection. In the Introduction to *Metaphysical Foundations*, Kant claims that a complete system of principles should be provided. In order to do that, he proposes to proceed exactly as he proceeded in the Analytic of Principles in obtaining the complete system of the pure principles of the understanding, that is, beginning with the four classes of categories: 'But the scheme for completeness of a metaphysical system, whether it be of nature in general, or of corporeal nature in particular, is the table of categories' (M, p. 10, 473).

This means analysing the concept of matter, characterizing the metaphysical level limited to the objects of the external senses, from the point of view of the four categories. In this way, four classes of principles are obtained which, then, allow us to have a mathematized discipline, and to construct the concepts of objects discussed by physics. Since the classes of categories are four, so there are four points of view from which to analyse the concept of matter. Each of these makes a particular chapter of physics possible: (1) *phoronomy*, (2) *dynamics*, (3) *mechanics* and (4) *phenomenology*.

Four classes of first metaphysical principles are also obtained. These specify respectively: (1) how to construct concepts of objects in motion in space, independently of the forces causing their motion; (2) how to construct concepts of objects in motion, considering the forces causing their motion; (3) how to construct concepts on the relations between the motions of objects, taking forces into account; (4) how to estimate in a modal sense the construction of concepts of objects in motion, taking forces into account.

Of these four classes, the most interesting is the third, since the principles of (Kantian) mechanics are exactly three laws of the Newtonian mechanics:[46] the principle of the conservation of matter (mass); the principle of inertia; the principle of action and reaction. Moreover, through a comparative observation, it seems that they could be seen as particularizations of the three analogies of experience to the objects of the external senses, that is, to material objects. However, it is worth mentioning that this particularization of one specific group of the pure principles to a specific group of metaphysical principles, on the grounds of implementing the empirical concept of matter, is something that can be determined only a posteriori, that is, only after the group of metaphysical principles has been found, exactly as the other three groups have been found. For that possible particularization applies only to the principles of mechanics, certainly not to those of phoronomy, nor to those of dynamics or of phenomenology.[47]

There are three aspects still to consider before concluding the analysis of the metaphysical level. The first concerns whether the metaphysical principles could be justified; the second regards the relation between pure physics and empirical physics; the third deals with the lawfulness of nature. As far as the first is concerned, it should be remembered that the justification of the necessity of the pure principles is, in the *Critique of Pure Reason*, indicated by *Beweis*: the 'transcendental proof', as Paton call it. *Beweis* is the same term Kant uses to indicate the justification of the first metaphysical principles, whose text is inserted within what he calls *Lehrsatz*. However, claiming that the justification of the metaphysical principles follows the same method and that it has the same status as the justification of the pure principles would be a mistake, particularly because the pure principles concern the possibility of the object 'in general', while the metaphysical principles concern the possibility of a mathematical physics. However, there is something similar, something which spurs me to speak in terms of *Beweisart*, to refer to a typical Kantian terminology (KdrV, p. 237, B264), which can be applied analogically to two different fields. In both fields it is demonstrated that without those principles a certain thing (nature 'in general', experience 'in general', object 'in general', in the case of the pure principles; the mathematical construction of the concepts of the physical objects, in the case of the metaphysical principles) could not be possible. Now, if we follow Paton and call 'transcendental justification' the justification of the first, the justification of the second might be called 'metaphysical justification', indicating that it is a *Beweis* showing that only that particular first metaphysical principle is a priori capable of taking into account that particular aspect of the matter 'in general' in motion.

Regarding the second aspect, it should be pointed out that Kant has never written nor claimed that all physics is a priori, that is, pure. Actually, it has a pure part made up of the first metaphysical principles. But any other physical principle, or any other physical law, including the law of universal gravitation (see M, pp. 56–7, 518), must be considered as an empirical law. With reference to this point (to which I will return later) note that such an empirical law *has been obtained by reflecting on experience,* thanks to the reflecting capacity of judgement, *and not inductively from experience.* From this point of view, it should not be embarrassing when Kant writes that: 'no law of either attractive or repulsive force may be risked on *a priori* conjectures. Rather, everything, even universal gravitation as the cause of the weight, must be inferred, together with its laws, from data of experience [*aus Datis der Erfarhung*]' (Ibid., p. 73, 534).

The last sentence might seem to contradict the position expressed in the *Critique of Pure Reason,* where Kant claims that knowledge does not begin 'out of' experience, but 'with' experience. It might seem, but it is not so. I am rather inclined to exclude altogether an inductivist interpretation of Kant's position, and, thus, the idea that here he has fallen into a contradiction. I read the quoted passage as a unhappy formulation, as a passage where the pen has moved faster than the mind.

Finally, the problem of lawfulness. If at the transcendental level the legislating understanding produces the lawfulness of nature 'in general', that is, produces *natura formaliter spectata,* at the metaphysical level something different happens.[48] We have seen that the first metaphysical principles make the mathematization of physics possible. That is, they allow us a particular way of representing both objects of the external senses and abstract objects (not to be confused with objects 'in general'). This means that they allow us a mathematical representation of the lawfulness of physical nature that, as particularization of nature 'in general', is made cognitively significant by the pure principles. In other words, at the metaphysical level, we have the lawfulness of a mathematized particular nature.

2.2.4 The empirical level

Judgements of perception and empirical judgements

Considering synthetic judgements a posteriori, Kant proposes a division on the basis of the fact that: 'judging can be of two types: first, when I merely compare the perceptions and connect them in a consciousness of my state, or, second, when I connect them in a consciousness in general' (P, p. 53, 300).[49] Therefore, there are judgements of perception, or perceptive judgements (*Wahrnehmungsurteile*), connected to the consciousness of an individual perceptive act, and judgements of experience (*Erfahrungsurteile*), connected to consciousness 'in general': the 'transcendental unity of apperception' (*die transzendentale Einheit der Apperzeption*) (cf. KdrV, p. 157, B139).[50] Kant explains that,

> although all judgments of experience are empirical, that is, have their basis in the immediate perception of the senses, nonetheless the reverse is not the case, that therefore all empirical judgments are judgments of experience; rather, beyond the empirical and in general beyond what is given in sensory intuition, special concepts must yet be added, which have their origin completely *a priori* in the pure understanding, under which every perception can first be subsumed and then, by means of the same concepts, transformed into experience. *Empirical judgments, insofar as they have objective* validity, are *judgments of experience*; those, however, that are *only subjectively valid* I call mere *judgments of perception*. The latter do not require pure concepts of the understanding, but only the logical connection of perception in a thinking subject. But the former always demand, beyond the representations of sensory intuition, in addition special *concepts originally generated in the understanding*, which are precisely what make the judgments of experience *objectively valid*.
>
> (P, p. 51, 297–8)

For example,

> that the room is warm, the sugar is sweet, the wormwood repugnant, are merely subjectively valid judgments. I do not at all require that I should find it so at every time, or that everyone else should find it is just as I do; they express only a relation of two sensations to the same subject, namely myself, and this only in my present state of perception, and are therefore not expected to be valid for the object: these I call judgments of perception. The case is completely different with judgments of experience. What experience teaches me

> under certain circumstances, it must teach me every time and teach everyone else as well, and its validity is not limited to the subject or its state at that time. Therefore I express all such judgments as objectively valid; as, for example, if I say: this air is elastic, then this judgment is to begin with only a judgment of perception; I relate two sensations in my senses only to one another. If I want it to be called a judgment of experience, then I require that this connection be subject to a condition that makes it universally valid.
>
> (Ibid., pp. 52–3, 299)

Therefore, Kant divides the singular judgements a posteriori (it is important to notice they are *singular* judgements) into (1) perceptive judgements, which are merely subjective because they simply connect two sensations of the same perceiving subject, and (2) judgements of experience, which are objective because this kind of connection falls under a certain condition which makes it universally valid on the basis of intervention of the understanding.

An individual perceiving subject, say S, can have perceptive judgements of the type 'a_S is b_S', where *a* and *b* are S's subjective sensations. If this judgement is to be objectified, it must have universal validity (Ibid., p. 51, 298). That is, it must become a judgement of experience, '*a* is *b*', through the universalization of the relation thanks to a particular impositions of the understanding. But how is this possible?

Before addressing this problem, two aspects should be noted; one explicitly pointed out by Kant himself, the other one implicitly contained in his works. The first regards the notion that only certain perceptive judgements can become judgements of experience; for there are perceptive judgements that never become judgements of experience:

> I gladly admit that these examples do not present judgments of perceptions such as could ever become judgments of experience if a concept of understanding were also added, because they refer merely to feeling – which everyone acknowledges to be merely subjective and which must therefore never be attributed to the object – and therefore can never become objective; I only wanted to give for now an example of a judgment that is merely subjectively valid and that contains in itself no basis for necessary universal validity and, thereby, for no relation to an object. An example of judgments of perception that become judgments of experience through the addition of a concept of the understanding follows in the next note.
>
> (Ibid., p. 52, fn *, 299)

And

> To have a more easily understood example, consider the following: If the sun shines on the stone, it becomes warm. This judgment is a mere judgment of perception and contains no necessity, however often I and others also have perceived this; the perceptions are only usually found so conjoined. But if I say: the sun *warms* the stone, then beyond the perception is added to the understanding's concept of cause, which connects *necessarily* the concept of sunshine with that of heat, and the synthetic judgment becomes necessarily universally valid, hence objective, and changes from perception into experience.
> (Ibid., p. 54, fn *, 301)

Therefore, from the point of view of the transition from subjective to objective, there are: (1) perceptive judgements that never become judgements of experience (for instance, 'I like warmth'); (2) perceptive judgements (for instance, 'When the Sun falls on a stone, I feel it become warm'), which can become judgements of experience (for instance, 'The Sun warms up a stone') by the application of a pure concept (in this specific case, the concept of 'cause').

The second aspect, implicitly present in Kant's writings and examples, concerns the fact that he is not referring to all the possible judgements of experience, but to a particular subclass; precisely to that class in which the binding between the subject and the predicate is connected with a universalization due to a concept belonging to the understanding, in particular that of cause. Actually there should also be judgements of experience in which the relation between the subject and the predicate, although being objective, it is not at all universalizable in a causal way. I am referring to judgements such as 'Kant is a philosopher', 'Venezia is in Italy and Venice in California', and so on. But, if we are alert, we may note that Kant discusses only *singular judgements which can be derived, through subalternation, from universal judgements*. This is important as far as the problem of the difference between accidental universals and nomological universals is concerned, that is, with reference to the question of lawness.

Taking these aspects into account, from the point of view of the universality of the relation between the subject and the predicate, among the judgements of experience are:

(1) singular judgements (not considered by Kant), which are about non-generalizable particular phenomena (for example, 'Kant is a philosopher');

(2) singular judgements (considered by Kant), which are about generalizable phenomena (for example, 'The Sun warms up a stone').

Let us come back to the initial problem. Certainly the judgements of experience are a posteriori, but in what sense are they characterized by universality, being contingent judgements?

> But how does this proposition: that judgments of experience are supposed to contain necessity in the synthesis of perceptions, square with my propositions, urged many times above: that experience, as *a posteriori* cognition, can provide merely contingent judgments?
>
> (Ibid., p. 58, fn *, 305)

As far as the universal-singular relation is concerned, the universality here considered by Kant is not, obviously, the universality intended from the point of view of the quantity of the judgements, but from the point of view of the intersubjective – and thus universal – validity of the singular judgement 'a is b'. On the contrary, as far as the necessity-contingency relation is concerned, the solution proposed by Kant is simply a repetition of the earlier position:

> If I say: experience teaches me something, I always mean only the perception that is in it – for example, that upon illumination of the stone by the sun, warmth always follows – and hence the proposition from experience is, so far, always contingent. That this warming follows necessarily from illumination by the sun is indeed contained in the judgment of experience (in virtue of the concept of cause), but I do not learn it from experience; rather, conversely, experience is first generated through this addition of a concept of the understanding (of cause) to the perception
>
> (Ibid.)

Actually the real solution should be sought in the extremely important and central §19 of the Transcendental Deduction, where Kant provides a new definition of judgement, and where a clue to the heart of his thought can be found.[51] The paragraph begins in a peremptory way:

> I have never been able to accept the interpretation which logicians give of judgment in general. It is, they declare, the representation of

> a relation between two concepts [...] I need only point out that the definition does not determine in what the asserted *relation* consists. But if we investigate more precisely the relation of a given mode of knowledge in any judgment, and distinguish it, as belonging to the understanding, from the relation according to laws of reproductive imagination, which has only subjectivity validity, I find that a judgment is nothing but the manner in which given modes of knowledge are brought to the objective unity of apperception. This is what I intended by the copula 'is'.
>
> (KdrV, pp. 158–9, B140–1)

Up to Kant, following the Aristotelian-medieval tradition, judgement had been considered as a *compositio* between two concepts: the subject-concept and the predicate-concept. Now the status of this *compositio* is questioned, and to find a way out the copula must be focused.[52] For it is exactly through the copula that we can distinguish the necessary and objective unity from the merely subjective connection, which, in this §19 and in the previous §18, Kant calls 'subjective unity', 'determination of the internal sense', 'association' and 'subjective state'.

Therefore, judgement must be thought of as 'the way to reduce given knowledge to the objective unity of apperception'. It is on the basis of this definition that the status of 'judgement' should be denied to what, in the *Prolegomena* and in the *Logic*, he considers as judgements of perception. In fact these are mere associations of subjective states of consciousness, due to the 'reproductive imagination', and do not have to be analysed by philosophy, but by psychology (Ibid., p. 165, B152).

Let us continue with the following, fundamental passage:

> It [the copula 'is'] is employed to distinguish the objective unity of given representations from the subjective. It indicates their relation to original apperception, and its *necessary unity*. It holds good even if the judgment is itself empirical, and therefore contingent, as, for example, in the judgment, 'Bodies are heavy'. I do not here assert that these representations *necessarily* belong *to one another* in the empirical intuition, but that they belong to one another *in virtue of the necessary unity* of apperception in the synthesis of intuitions, that is, according to principles of the objective determination of all representations, in so far as knowledge can be acquired by means of these representations – principles which are all derived from the fundamental principle of the transcendental unity of apperception. Only in this way does there arise from this relation a *judgment*, that is, a

> relation which is *objectively valid*, and so can be adequately distinguished from a relation of the same representations that would have only subjective validity – as when they are connected according to laws of association. In the latter case, all that I could say would be, 'If I support a body, I feel an impression of weight'; I could not say, 'It, the body, is heavy'. Thus to say 'The body is heavy' is not merely to state that the two representations have always been conjoined in my perception, however often that perception be repeated; what we are asserting is that they are combined *in the* object, no matter what the state of the subject may be.
>
> (Ibid., p. 159, B142)

Therefore, empirical judgements are contingent, but they connect the representations in a universal and objective way thanks to the intervention of the categories of relation. Whenever Kant speaks about their 'necessity', he does not intend a necessity expressed by the proposition itself, but by the principle of the pure understanding (especially by the analogy of experience) thinkable of as a mould into which the representations are inserted so as to transform their mere subjective association into an objective union.

Therefore, two different degrees of objectivity should be differentiated:

(1) *strong objectivity*, typical of the pure principles of the understanding, that is, of the synthetic a priori judgements;
(2) *weak objectivity*, typical of empirical judgements, and thinkable of as a consequence of strong objectivity.

This could be a solution to our problem. There are associations of subjective states of consciousness that, as in the *Prolegomena* and in the *Logic*, we can call perceptive judgements, although we know it would not be correct. Some of these associations, in particular those not connecting mere subjective sensations, can be transformed, thanks to the intervention of the pure concepts, into empirical judgements (or judgements of experience). These latter, although being intrinsically contingent, are characterized by a weak objectivity, that is, they partake in the universality and necessity of those synthetic a priori judgements that have permitted them to be empirical judgements. On these grounds, 'I feel that this stone, lit up by the Sun, is warm' is a mere association of individual states of consciousness (of the type 'a_S is b_S', where S is the given perceiving subject) which, nevertheless and differently from 'I feel that this cake is sweet' (which is still of the type 'a_S is b_S'), can become an empirical judgement of the kind 'The Sun warms up a stone' (which is of the type '*a* is *b*'). It suffices to insert the representations 'This stone is lit up by the Sun' and 'This stone is warm' into the mould allowed by the principle of the second analogy of experience, that is, the pure principle connected with the category of causality.

There is a further point to consider. At the end of the quoted passage, Kant claims that *a repetition of a certain connection is not enough to make it* objective, *not even in the sense of weak objectivity. Instead the presence of an a priori mould is always necessary*. This observation is particularly important. Kant clearly abandons, as we are going to see in a while, Hume's position according to which it is exactly the repetition of a conjunction that gives it validity (even if only psychological validity). In Kant's view, the repetition of the connection does not give it any objective validity. It becomes objectively valid only if it is inserted into the objective mould offered by the understanding, that is, by the pure principles, which 'are all derived from the fundamental principle of the transcendental unity of apperception' (Ibid.).

Let us stay with the extremely important §19 of the Transcendental Deduction. It is the judgement, as we have seen, which makes the objective unity of apperception, that is, it is the judgement where the subjective association is objectivized on the basis of the understanding and the original apperception, the 'I think'. It should be recalled that knowing, especially knowing an object, means bringing it under a concept, that is, to predicate that object of that concept or, in other terms, to express a judgement; because the concept is a predicate of a possible judgement. According to Kant's position, between concepts and judgements, judgements come first. In other words, to have a concept of an object, I must represent this object to myself as something, which means to express a judgement on it.

The empirical judgement, objective judgement, is as it is because of the intervention of the understanding through its categories. Empirical judgement is where the two sources of human knowledge, sensibility and understanding, correspond to make possible empirical knowledge of the object in question. This is one of the many manifestations of the Kantian 'Copernican revolution', that is, if you like, it is the core of the theory-ladenness.[53]

Regarding the theory-ladenness of the given datum, it should be emphasized that not even at the perceptive level, which is nevertheless entirely subjective, is the knowing subject able to arrive at the bare given datum. Here also, although in a non-objectivizing form, the understanding intervenes, as one can easily infer from part of §18 of the Transcendental Deduction:

> the pure form of intuition in time, merely as intuition in general, which contains a given manifold, is subject to the original unity of consciousness, simply through the necessary relation of the manifold of the intuition to the one 'I think', and so through the pure synthesis of understanding which is the *a priori* underlying ground of empirical synthesis. Only the original unity is objectively valid; the empirical unity of apperception [to be read as 'perception'], upon which we are not dwelling, and which besides is merely derived from the former under given conditions *in concreto*, has only subjective validity.
>
> (Ibid., p. 158, B140)

In this passage, not only does Kant confirm the difference between the objectively valid unity, traceable to the 'I think', and the subjectively valid unity, traceable to the perceptive association, but also firmly declares that the latter has the former as its very basis, that is, that the latter 'only under given conditions derives *in concreto* from the first'. In short, the empirical association, that is, the subjective connection of representations, could not occur unless, in its foundations, there were the true and proper transcendental unity (see Scaravelli, 1968, p. 272). Unfortunately Kant concludes that he is not going to analyse specifically how all this really happens.

However this is not the whole story. Note that perceptive judgements must also possess a certain primitive objectivity. Certainly this does not characterize the connection between the subject-concept (for example 'a stone') and the predicate-concept (for example 'warm'), which, as emphasized, is subjective. But it characterizes the concepts ('warm', and 'stone) themselves. Already here the categories have been imposed by the understanding, in particular the categories of quantity and quality: there would be no other way of knowing that that thing is a 'stone' and that that thing has a certain 'grade of warmth'.

2.2.5 The empirical laws and the second analogy of experience

We know that all the phenomena, *in order to be given*, have to fall under the pure forms of intuition (space and time), and, *in order to be known*, they have to fall under the categories of the understanding, that is, under the pure principles of the understanding or, in other words, under the laws of nature 'in general'. However, apart from nature 'in general', which is constituted and regulated by the pure principles of the understanding (we are at the transcendental level), there is also nature in particular, which is also constituted and regulated by the empirical laws (we are at the empirical level):

> Special laws [*besondere Gesetze*, also called *empirische Gesetze* or *empirische Grundsätze* (see KdrV, pp. 194–6, B197–200)], as concerning those appearances which are empirically determined, cannot in their specific character be *derived* from the categories, although they are one and all subject to them. To obtain any knowledge whatsoever of these special laws, we must resort to experience; but it is the *a priori laws* that alone can instruct us in regard to experience in general, and as to what it is that can be known as an object of experience.
>
> (Ibid., p. 173, B165)

Now, the empirical laws:

(1) Are obviously different from the pure principles of the understanding which apply to nature 'in general' and have the transcendental function both of constituting *natura materialiter spectata*, and to regulate *natura formaliter spectata* (see also P, pp. 71–4, 318–20).
(2) Cannot be derived from the pure principles of the understanding, but they can be found by resorting to experience: 'empirical laws can exist and be discovered only through experience' (KdrV, p. 237, B263; see also KdU, pp. 23–4, 184–5 – II Introduction), even if, of course, not inductively.
(3) Although they cannot be established only on the basis of the pure principles of the understanding, have to submit to them, since such principles are the transcendental laws of nature 'in general', that is, of nature of which they – the empirical laws – rule the particular relations: 'these principles alone [the pure principles of the understanding] supply the concept which contains the condition, and as it were the exponent, of a rule in general. What experience gives is the instance which stands under the rule' (KdrV, p. 195, B198). This means that not only is there a nature 'in general', but also a nature 'in an empirical sense' (Ibid., p. 237, B263) particularizing the former, and that it is possible only thanks to the former.

At this point, at least two clarifications should be made. First, I must clarify the relation between the empirical laws and the pure principles, in particular the second analogy of experience regulating the determination of the possibility of the non-synchronous relations among phenomena. Second, I must clarify how we arrive at the empirical laws, taking into account the fact that the assumption of the 'Copernican revolution' must not be infringed, but that, at the same time, we must resort to experience to discover them.

First, let us put aside the huge secondary literature on the relation between the second analogy of experience and the empirical laws, and let us approach Kant's texts directly.[54] To begin with, it is worth recalling what the principle of the analogies of experience claims: 'experience is possible only through the representation of a necessary connection of perceptions' (Ibid., p. 208, B218). Now, as it is stated in the *Beweis* of this principle, experience is empirical knowledge (and thus objective knowledge) which, in the case of the connections of perceptions, is made possible precisely because of the analogies of experience. These regulate the connections among perceptions, in functions of the three modalities of time: permanence, succession and coexistence. It should be noted, as Kant himself emphasizes, that the analogies of experience concern neither the phenomena, nor the synthesis of their intuition. Phenomena have to do with the axioms of intuition and the anticipations of perception, which make possible 'the *existence* of such appearances and their *relation* to one another in respect of their existence' (Ibid., pp. 210–11, B222).

Therefore, first the phenomena have already been constituted as such. Then to objectively rule the relations among them, that is, to objectively rule the connections among the perceptions to which they are related, the analogies of experience intervene, that is, the regulative principles of the understanding, which 'will be prior to all experience, and indeed make it possible' (Ibid., p. 209, B219). Thus, there is a regularity of nature 'in general' made possible exactly by the analogies of experience. In other words, the legislating understanding, imposing the analogies of experience, which are regulative, that is, providing rules, constitutes the lawfulness of nature.[55] Of the three analogies, the second is particularly important, and it concerns the succession of perceptions in time, and thus their causal connection. With reference to causality, it should be recalled that one thing is the *category of causality*, another one is the *schema of causality*, and still another one is the *principle of causality*.

But what is the category of causality? In the previous chapter we saw that even if inside the Kantian approach a real definition of any individual category is impossible, a transcendental definition and an operational definition are possible. As far as the category of causality is concerned, we have an explicit operational definition:

> The given intuition must be subsumed under a concept, which determines the form of judging in general with respect to the intuition, connects the empirical consciousness of the latter in a consciousness in general, and thereby furnishes empirical judgments with universal validity; a concept of this kind is a pure *a priori* concept of the understanding, which does nothing but simply determine for an intuition the mode in general in which it can serve for judging. The concept of cause being such a concept, it therefore determines the intuition which is subsumed under it, for example, that of air [Kant is discussing how 'the air is elastic' is transformed from judgment of perception into judgment of experience], with respect to judging in general – namely, so that the concept of air serves, with respect of the expansion, in the relation of the antecedent to the consequent in a hypothetical judgment. The concept of cause is therefore a pure concept of understanding, which is completely distinct from all possible perception, and serves only, with respect to judging in general, to determine that representation which is contained under it and so to make possible a universally valid judgment. Now before a judgment of experience can arise from a judgment of perception, it is first required: that the perception be subsumed under a concept of the understanding of this kind; that is, the air belongs under the concept of cause, which determines the judgment about the air as hypothetical with respect to expansion. This expansion is represented not as belonging merely to my perception of the air in my state of perception or in several of my states or in the state of others, but as *necessarily* belonging to it, and the judgment: the air is elastic, becomes universally valid and thereby determines the perceptions not merely with respect to each other in my subject, but with respect to the form of judging in general (here, the hypothetical), and in this way makes the empirical judgment universally valid.
>
> (P, pp. 53–4, 300–1)

Or, more abstractly,

> [...] the concept of cause, which signifies a special kind of synthesis, whereby upon something, A, there is posited something quite different, B, according to a rule [...in a way such that the latter]

> follows from it *necessarily and in accordance with an absolutely universal rule.*
>
> (KdrV, pp. 124–5, B122–4)

Having specified the category of causality, we can move to (1) the schema of causality: 'The schema of cause, and of causality of a thing in general, is the real upon which, whenever posited, something else follows. It consists, therefore, in the succession of the manifold, in so far as that succession is subject to a rule' (Ibid., pp. 185, B183); and (2) to the principle of the second analogy of experience, that is, to the synthetic a priori judgement expressing the principle of causality: 'All alterations take place in conformity with the law of the connection of cause and effect' (Ibid., p. 218, B232).

Regarding the operative definition of the category of causality and the principle of the second analogy, it is worth noting that since the pure principles of the understanding indicate, through the schemata, how to apply the categories, then the operative definition of the category of causality is nothing but what is indicated by the corresponding pure principle. Generalizing, we can claim to have an operative definition of a category only if, and when, we know the corresponding pure principle applying it.

As we have seen in the previous passages, each time that a change of a certain thing 'in general' occurs, it is regulated by a necessary and universal law, that is, by the 'law of connection of cause and effect': the *principle of causality*. In other words, each time that an alteration occurs from a phenomenic situation A 'in general' to a phenomenic situation B 'in general', there is a universal and necessary judgment which rules it:

For each alteration leading to an event B 'in general',
there is an event
A 'in general'
such as
A 'in general' is the cause of B 'in general'

If we indicate 'A "in general"' by **A** and 'B "in general"' by **B**, then the *principle of transcendental causality*, contained in the principle of the second analogy of experience, can be formulated as

$$\forall x \mathbf{A}x \stackrel{C^T}{\rightarrow} \mathbf{B}x$$

where C^T indicates that we are dealing with causality in the transcendental sense. It is important to point out that **A** and **B** are neither event-type, nor tokens of event-type, nor particular events. They are events 'in general', that is, events which can be understood only within Kant's transcendental architectonic.

The interpretation of the second analogy I am proposing is equivalent neither to that according to which 'each event has a cause', nor especially to that according to which 'the same cause produces the same effect'. The first interpretation is denied on the basis of the fact that the principle of the second analogy is not simply the principle of causality, but the principle according to which each change 'in general' falls under the principle of causality. Thus, it does not follow that each event has a cause, but each event 'in general' which concludes (also temporally) a change process has a cause 'in general'. As I will show, only in this way is it then possible to accept non-causal (and in this case also non-'in general') explanations of empirical phenomena. The second interpretation is denied as well, since claiming that the same causes have the same effects means claiming that, on the one hand, nature is uniform and, on the other hand, that there is a particular causal law for those event-types, but this is not at all contained in the principle of the second analogy.

In brief, here we are dealing with A and B 'in general'; we are at the transcendental level where nature, even if *formaliter spectata*, has to be intended 'in general'. And, obviously at this level, the problem of lawness is trivial: the universal statement $\forall x \mathbf{A}x \overset{C^T}{\rightarrow} \mathbf{B}x$ is certainly a law, since produced by the legislating understanding.

I will return again to this point, but now I wish to focus on some aspects of the *Beweis* of the principle of the second analogy supporting my interpretation. Perceiving two phenomena occurring one after the other means connecting them in time. However, this succession of perceptions is subjective, and it is a product of the reproductive imagination. In order to have an objective succession, I need a pure concept, in particular that 'of the relation of cause and effect, the former of which determines the latter in time, as its consequences – not as in a sequence that may occur solely in the imagination' (Ibid., p. 219, B234).

Thus, as we have seen in the case of the transition from perceptive 'judgements' to judgements of experience, not only is a certain succession made objective, but its experience is also made possible, or 'the appearances, as objects of experience, are themselves possible only in conformity with the law' (Ibid.).

Therefore, if I want the perception of a change to have a cognitive sense, such a change has to fall under the principle of the second analogy of experience, according to which all changes are causal processes.

However, this must not be intended in the sense that the events concerning things in themselves (those affecting the senses in relation to the corresponding perceptions) are bound by causal relationships. Instead it must be intended 'in general', in the sense that the representations of two successive states are bound by a causal relation (Ibid., pp. 219–20, B234–6).

Let us analyse two of Kant's examples:

(1) I look at the ship travelling down the river; the perception B of its being in a certain position, follows the perception A of its being in a previous position;
(2) I look at a house; the perception B of one part, for instance, the roof, follows the perception A of another part, for instance, its foundation.

In both cases, there is a succession of perceptions. But are they both objective successions, that is, causal successions?

In the first case, 'B can be apprehended only as following upon A; the perception A cannot follow upon B but only precede it' (Ibid., p. 221, B237). However, in the second case, B can precede A, and vice versa. Thus, there is something different, in particular, in the former an irreversible succession is described, while in the latter a reversible one is described. The reason for this difference is that in the first case the subjective succession corresponds to the objective order, that is, that objectivized as a result of the causal rule, while in the second case, the subjective succession does not correspond to any objective order; even if it could still be objectivized in a different way thanks to a different rule: the rule related to the principle of the third analogy, that is, concerning simultaneousness.

Therefore, a causal relation is characterized not only by the fact that the phenomena are correlated *temporally*, but also that there is an *irreversible* time-asymmetry linked to their causal asymmetry. Moreover, since only the causal relation allows us experience 'in general', and since its rule enters a synthetic a priori judgement, which is apodictic, that is, atensionally true, then (1) a first occurring event remains a first occurring event for ever; and (2) a temporal consequent event remains a temporal consequent event for ever. In other words, what is a cause is a cause for ever, and what is an effect is an effect for ever; this relation is *temporally invariable*.

There is more to add. Let us return to the example of the ship. It is first perceived as A and then as B. However, this does not mean that A causes B, even if we are dealing with an objective succession, and therefore with a causal succession. In fact that A precedes B is causally determined in the sense that there are forces that necessarily cause it first to be in a given position and then to be in the successive position. This implies that in order to speak properly of cause and effect, there must be a *spatial contiguity* between what is cause (the river flowing from a given point to a successive one) and what is effect (changing position of ship's keel from the first point to the second one). With reference to this, it should further be noted that the causal explanation of the ship travelling down the river is analogous to the explanation of another phenomenon that had troubled philosophers, that is, the continuous succession of night and day. Is this a causal succession, knowing that night does not cause day, nor day cause night? In fact, night and day are effect-phenomena which necessarily follow the cause phenomenon: the reciprocal motion of the Earth and the Sun.[56]

Let us return to the transcendental causal relation, that is, to the second analogy, and summarize what we found: it is an irreversible temporal relation between two spatially contiguous events 'in general'. However, this does not imply that we know how the particular causal relation is at the empirical level, and thus neither do we know if nature is uniform, that is, if the 'same cause means the same effect'.

To take another example from Kant, we know that to place a ball on a cushion produces a concavity. In this case, there is no apparent temporal succession between the phenomena in question, but apparent simultaneity. However, as Kant points out, in the case of the causal relation we should not look for the *temporal course*, which can also tend to zero, as in the case just seen, but for the *temporal order*: 'the time between the causality of the cause and its immediate effect may be [a] *vanishing* [quantity], and they may thus be simultaneous; but the relation of the one to the other will always still remain determinable in time' (Ibid., p. 228, B248).

In other words, even if the phenomena in question are apparently simultaneous, before placing the ball, the cushion was flat and only by placing the ball (related to the cause) is the concavity produced (the effect). In this case, the course between the placing of the ball and the formation of the concavity is approximately zero, but there is a temporal order between what is the cause and what is the effect.

Once this point is clear, we can turn to another relevant point; that concerning the relation between the particular causal laws and transcendental causality. We have already seen that in order to perceive a change not merely as a subjective succession, but as a temporal objective succession, a rule establishing what comes first and what comes after must intervene. We also know that this rule is a priori. *However, we do not know anything else, or, rather, we do not know anything else at this a priori level.*

We should not be deceived by the examples discussed by Kant: the Sun warming up a stone, the Sun melting wax, the river pushing a ship, the weight of the ball producing a concavity and so on. In all these cases, we know which the causal force in question is, and, thus, we could interpret the second analogy as affirming that not only is the succession of perceptions objectivized thanks to the causal rule, but also that all similar cases fall under the same rule. However, this is not true. As said before, Kant is absolutely clear in that respect: the particular causal empirical laws cannot be known a priori, nor deduced from the principles a priori, especially from the principle of the second analogy. In fact

> how anything can be altered, and how it should be possible that upon one state in a given moment an opposite state may follow in the next moment – of this we have not, *a priori*, the least conception. For that we require knowledge of actual forces, which can only be given empirically, as, for instance, of the moving forces, or what amounts to the same thing, of certain successive appearances, as motions, which indicate [the presence] of such forces. But apart from all question of what the content of the alteration, that is, what the state which is altered, may be, the form of any alteration, the condition under which, as a coming to be of another state, it can alone take place, and so the succession of the states themselves (the happening), can still be considered *a priori* according to the law of causality and the conditions of time.
>
> (Ibid., pp. 230, B252)

Only the formal condition of the possibility of knowledge of any alteration is given a priori, that is, we know a priori only that each time an alteration occurs, $\forall x \mathbf{A}x \overset{C^T}{\rightarrow} \mathbf{B}x$ holds, but certainly we do not know the particular instantiation of the causal law underlying that particular alteration. Only by *reflecting* (we will see that this verb is not fortuitous) on the particular phenomenic situation, can the knowing subject arrive at the determination of the universal statement causally connecting an A-type event to a B-type event, that is, to a statement such as

$$\forall x Ax \overset{C^E}{\rightarrow} Bx.$$

Now the causal relation between being A and being B is no longer a transcendental relation (C^T), but an empirical relation (C^E). However, at this point, two new problems arise: (1) how can we arrive at the formulation of $\forall x Ax \overset{C^E}{\rightarrow} Bx$? (the problem of the logic of discovery); (2) how do we know that the universal conditional statement which we produce ($\forall x Ax \rightarrow Bx$) is really a universal conditional statement of the type $\forall x Ax \overset{C^E}{\rightarrow} Bx$, that is, a law? (Schlick's problem, or the problem of lawness.) These two problems are, as I will show, strictly bound up together in the Kantian approach. However, before moving to them, I should consider, now that we have all the necessary elements, how Kant solves *Hume's challenge* on causality.

We know the essence of Hume's criticism of causality, and Kant seems to accept it completely, as can be inferred from the *Beweis* of the principle of the second analogy of experience and other passages of the *Critique of Pure Reason* (for example, p. 44, B4–5; p. 124, B122–3; and pp. 605–12, B786–97). Nevertheless Kant does not accept the psychological solution, based on the habit of associations, proposed by Hume. Instead he interprets the principle of causality as an a priori rule establishing the objectivization of temporal succession, and, thus, contributing to regulating the possibility of experience 'in general', that is, to the lawfulness of nature 'in general', or to the determination of *natura formaliter spectata*.[57]

However, what motivates Kant to deny Hume's solution? The reasons for this are considered at length in Chapter II of the Transcendental Doctrine of Method, where we can also find a further argument in favour of the reading of the principle of the second analogy that I have presented. For Kant, Hume, although being 'the most ingenious of all the skeptics' (Ibid., p. 609, B792), made an extremely serious mistake: he passed from the impossibility of obtaining a necessary and universal judgement from a contingent and singular empirical judgement, to the impossibility of having a necessary and universal judgement at all;

> That sunlight should melt wax and yet also harden clay, no understanding, he pointed out, can discover from the concepts which we previously possessed of these things, much less infer them according to a law. Only experience is able to teach us such a law. But, as we have discovered in the Transcendental Logic, although we can never pass *immediately* beyond the content of the concept which is given us, we are nevertheless able, in relation to a third thing, namely, *possible* experience, to know the law of its connection with other

> things, and to do so in *a priori* manner. If, therefore, wax, which was formerly hard, melts, I can know *a priori* that *something* must have preceded, ([that something being] for instance [in this case] the heat of the sun), upon which the melting has followed according to a fixed law, although *a priori*, independently of experience, I could not determine, *in any specific manner*, either the cause from the effect, or the effect from the cause. Hume was therefore in error in inferring from the contingency of our determination *in accordance with the law* the contingency of the *law* itself. The passing beyond the concept of a thing to possible experience (which takes place *a priori* and constitutes the objective reality of the concept) he confounded with the synthesis of the objects of actual experience, which is always empirical [...], which exists only in the imitative faculty of imagination, and which can exhibit only contingent, not objective, connection.
>
> (Ibid., pp. 610–11, B794)

Here Hume mixed up, as Kant sees it, real (physical) contingency/necessity with logical contingency/necessity, and with transcendental contingency/necessity. That the particular laws concern physically contingent phenomena implies neither the impossibility of the physical necessity (characterizing causality at the empirical level), nor that this latter is based, in the sense of the transcendental logic and not in the sense of the formal logic, on transcendental necessity (characterizing causality at the transcendental level).[58]

2.2.6 The unity of system

Although most commentators on Kant agree that the Analytic of Principles does not solve the problem of the empirical laws, some maintain that it is solved in the Appendix to the Transcendental Dialectic of the *Critique of Pure Reason*, while others affirm it is solved in the *Critique of Judgement*.[59]

Those who argue for the first possibility base their interpretations on the consideration that the Appendix anticipates almost all of what is going to be at issue in the *Critique of Judgement*. As far as this interpretative proposal is regarded, I am inclined to consider it as forcing too much into Kant's words, especially if we wish to save the architectonics of his critical philosophy, theorized in the Doctrine of Method of the *Critique of Pure Reason*; practically displayed with the three *Critiques*; particularized in the *Metaphysical Foundations of Natural Science*, as far as the topic on nature is concerned ('the starry heaven above me'), and in *Metaphysics of Morals*, as far as the problem of freedom is concerned ('the moral law within me'). It should also be noted that Kant himself writes explicitly that the critique of pure reason without the critique of capacity of judgement would be incomplete (KdU, pp. 4–5, 167–8 – Preface).

However, the thesis of the anticipation in the Appendix of topics analysed in the third critique is problematic, especially if we make a comparative reading of the two writings. In this way, not only do we find, as pointed out by Allison (1994, p. 305, n. 5), that some commentators make a mistake in failing to distinguish clearly between reason in its regulative use and the reflecting capacity of judgement, but also in neglecting other important differences. For example,

(1) The problem of the Appendix concerns the government of the unconditioned, with which Kant copes by interpreting regulatively the principles of reason; instead, as pointed out by Cassirer (1918, ch. VI, §2), the problem of the *Critique of Judgement* regards the relation between the (nomological) universal and the particular. Consequently, the Appendix, on the grounds of the solution of the unconditioned in regulative terms, focuses on the idea of unity of nature and the concept of system of laws; while in the third *Critique* the way in which the empirical particular is grasped nomologically by (either causal or teleological) laws is thematized.
(2) In the Appendix, even if the topic of teleological laws is also faced, there is no important distinction between the internal and the external purposiveness, which is discussed at length in the third *Critique*.
(3) In the Appendix, purposiveness is strictly connected to the third transcendental idea, while in the *Critique of Judgement* it is introduced as an 'heautonomous' principle of the reflecting capacity of judgement.
(4) Obviously, in the Appendix there is no mention of either the reflecting capacity of judgement or the aesthetical issues.

I should consider these points one by one. But first the problems of unity of nature and of systemacy of knowledge, discussed in the Appendix, must be analysed. Then I will turn to the third *Critique* and, thus, to the problem of the relations between the universal and the particular, that is, to the problem of the formulation of the (either causal, or teleological) empirical laws. I have already observed that to find an unconditioned condition 'whereby its [of the understanding] unity is brought to completion' (KdrV, p. 306, B364) is a 'requirement of reason' (*Forderung der Vernunft*); which in the Appendix is called 'interest of reason' (*Interesse* der Vernunft) in its logical use. We know that the unconditioned can be found by reasoning prosyllogistically on the basis of the three forms of the syllogism: the categorical, the hypothetical and the disjunctive syllogism. In that way, the three transcendental ideas can be identified. If they are considered objectively, they lead to dialectical conclusions, while, if considered as '*focus imaginarius*', that is, as a regulative ideal, they are what Kant regards as 'an excellent, and indeed indispensably necessary, regulative employment' (Ibid., p. 533, B672).

It is from this position we should start reading the Appendix. For what does it imply to have a *focus imaginarius*? Nothing but to have something, in particular an idea, focusing, not in the objective sense (that is, not in the sense of something real), but in the subjective sense (that is, in the regulative sense), what we know by means of the categories. Note that 'focus' should be intended exactly in the optical sense, that is, the point on which the light rays converge. Analogously, the transcendental idea allows us the convergence of what is already constituted by the imposition of the categorial apparatus, that is, what is already known. This means that the transcendental idea allows us the *unity of knowledge*. It allows us to have a set of synthetic judgements structured as a *system*: 'This idea accordingly postulates a complete unity in the knowledge obtained by understanding, by which this knowledge is to be a not mere contingent aggregate, but a system connected according to necessary laws' (Ibid., p. 534, B673).[60]

Therefore, 'the systematic unity of the manifold knowledge of the understanding [...] is a *logical* principle [...which assists the understanding] in those cases in which the understanding cannot by itself establish rules' (Ibid., p. 535, B676). And this occurs because 'the human reason has a natural tendency to transgress' the limits of possible experience (Ibid., p. 532, B670). Naturally, one must be careful to use this rigorously in a regulative way.

In short, understanding unifies (from the transcendental point of view) the empirical multiplicity through categories; reason unifies (from the regulative, that is, methodological, point of view) the 'manifold of concepts by means of ideas' (Ibid., p. 533, B672). Moreover since concepts are predicates of possible judgements, reason also unifies in a system the empirical judgements, or the empirical laws, that is, what Scaravelli (1968, pp. 357–68) rightly calls 'the third multiplicity'.[61] Therefore, at the cognitive top, that is, at the transcendental level, we have the eight pure principles of the understanding (plus the pure principles derived from them), which make nature 'in general' possible. At the metaphysical level, we have the twelve first metaphysical principles allowing us to have a mathematized and almost-axiomatized discipline of physical objects (those falling under the external senses). At the empirical level, we have both the infinite universal empirical judgements, that is, the infinite empirical laws, and the infinite singular and particular empirical judgements, most of which are unified in a system.

> On the other hand, since the laws that pure understanding gives *a priori* concern only the possibility of a nature as such [*einer Natur überhaupt*[62]] (as object of sense), there are such diverse forms of nature, so many modifications as it were of the universal transcendental concepts of nature, which are left undetermined by these laws, that surely there must be laws for these forms too. Since these laws are empirical, they may indeed be contingent as far as *our* understanding can see; still, if they are to be called laws (as the concept of a nature does require), then they must be regarded as necessary by virtue of some principle of the unity of what is diverse, even though we do not know this.
>
> (KdU, pp. 19, 179–80 – II Introduction)

This infinite multiplicity of empirical laws, which are contingent in comparison with the products of the pure understanding *but necessary exactly because they are laws,*[63] is unified in a system. Such a system is made possible only thanks to reason used regulatively. It is reason that seeks the unconditioned in the conditioned series.

At this point the matter is not yet totally clear. We need to précis some more of the ideas allowing us such unity, that is, the principles guiding the formation of the system of laws. Then, we should analyse the epistemological status of these principles and, thus, justify them. As far as the first problem is concerned:

> Reason thus prepares the field for the understanding: (1) through a principle of *homogeneity* of the manifold under higher genera; (2) through a principle of the *variety* of the homogeneous under lower species; and (3) in order to complete the systematic unity, a further law, that of *affinity* of all concepts – a law which prescribes that we proceed from each species to every other by gradual increase of the diversity.
>
> (KdrV, p. 542, B685–6)

We can call these principles, as Kant suggests:

(1) the principle of homogeneity (‘*entia praeter necessitatem non esse multiplicanda*’); which is related to *the idea of unity*;
(2) the principle of specification (‘*entium varietates non temere esse minuendas*’); which is related to *the idea of multiplicity*;
(3) the principle of continuity (‘*continui specierum – formarum logicarum*’); which is related to *the idea of affinity*.

Strictly speaking, only the first principle, that concerning the subsumption of species to genus, and of genus to higher genus on the basis of homogeneity, which implies the use of Ockham's razor, is clearly related to reason's tendency to the unity. Nevertheless, the second principle is necessary to limit the ‘possible indiscretion in the former principle’ (Ibid., p. 540, B682), that is, the ‘tendency towards unity’ (Ibid., p. 543, B688). For it is necessary to focus our attention on the fact that if a Porphyry-tree, relatively to concepts, can be run bottom-up, that is, from what is more particular to what is more general, it can be also run top-down, that is, towards the less general. In the first case, we find the system; in the second we find the chaotic infinite multiplicity of particular empirical laws and singular empirical judgements. Finally, the third principle, on the one hand, fulfils Kant's architectonic aim in its tripartite arrangement, and on the other hand, allows us the connection between the first two principles, since its invitation is to seek for the affinity which has to be found because species ‘all spring from one highest genus, through all degrees of a more widely extended determination’ (Ibid., p. 542, B686).[64] Now we need to understand better what these principles state, and what their status is. First, Kant gives them a name: ‘I entitle all subjective principles which are derived, not from the constitution of an object but from the interest of reason in respect of certain possible perfection of the knowledge of object, *maxims* of reason’ (Ibid., p. 547, B694).

Both in the Appendix and in the *Critique of Judgement*, there are passages where the status of the maxims (principles) of reason and the reflecting capacity of judgement is discussed. Nevertheless Kant seems to offer many contradictory theses, and therefore some commentators (for instance, Kemp Smith, 1923, p. 547) have strongly questioned the consistency of the Prussian philosopher, in particular in the Analytic. For running through the pages of the Analytic and those of the *Critique of Judgement*, we can find that the maxims, or the principles, or the related ideas:

(1) are not yielded out of nature (Ibid., p. 534, B673);
(2) are not constitutive (Ibid., p. 535, B675);
(3) are transcendental, and thus make systematic unity necessary, not only considered subjectively and logically as a method, but also considered objectively (Ibid., p. 536, B676; KdU, pp. 18–19, 179–180 – II Introduction);
(4) are regulative (KdrV, p. 547, B694; KdU, p. 259, A379);[65]
(5) have objective, even if indeterminate, reality (KdrV, p. 547, B693; KdU, pp. 287–8, 404) (KdrV, p. 549, B697);
(6) are subjective (Ibid., p. 305, B362, and p. 547, B694; KdU, pp. 23–4, 184);
(7) are 'as if' (for example, KdrV, p. 550, B699);
(8) are leading threads (for example, KdU, p. 259, 379).

Only at a first look are there contradictions among some of these eight points (see O'Shea, 1997, pp. 229–37). It should be noted that when Kant speaks about the subjective principles, he intends to oppose them to the possibility of objectivizing them in the sense of the possibility of using reason dialectically. However, they are transcendental principles allowing us to consider the unity of nature *as if* it was objective. Moreover, they have to possess 'a certain objective value': only in that case can the systemacy (not accomplished by understanding alone) be reached and, thus, *the interest of reason* can be fulfilled. With reference to this aspect, note that they do not derive from nature, but rather from such an interest of reason.

> The unity of reason is the unity of system; and this systematic unity does not serve objectively as a principle that extends the application of reason to objects, but subjectively as a maxim that extends its application to all possible empirical knowledge of objects. Nevertheless, since the systematic connection which reason can give to the empirical employment of the understanding not only furthers its extension, but also guarantees its correctness, the principle of such systematic unity is so far also objective, but in an indeterminate manner (*principium vagum*).
>
> (KdrV, p. 556, B708)

Therefore, we should understand in what sense the maxims are, on the one hand, objective and, on the other hand, subjective.[66] But another aspect is also worth observing, that concerning indeterminacy. This is made more explicit in an earlier passage where indeterminacy is related with the objectivity (actually, with a particular meaning of objectivity) of the maxims:

> the *unity of reason* is in itself *undetermined*, as regards the conditions under which, and the extent to which, the understanding ought to combine its concepts in systematic fashion [...] the principles of pure reason must also have objective reality in respect to that object, not, however, in order to *determine* anything in it, but only in order to indicate the procedure whereby the empirical and determinate employment of the understanding can be brought into complete harmony with itself. This is achieved by bringing its employment, so far as may be possible, into connection with the principle of thoroughgoing unity, and by determining its procedure in the light of this principle.
>
> (Ibid., pp. 546–7, B693–4)

The maxims are *as-ifs* having objective validity, but not the kind of objectivity that their dialectic use would lead to (on the contrary, as far as this fallacious possibility is considered they are to be considered only subjectively).[67] They are guiding principles, that is, a 'guide for reflection' (KdU, p. 269, 389).[68] Eventually they are the transcendental principles, and as such they necessitate a transcendental deduction (justification), since they have a 'certain objective value'. To provide it means to 'complete the critical work of pure reason' (Ibid., p. 549, B698), even if, as Kant emphasizes, their deduction is completely different from the deduction of the categories. In fact Kant provides not only a transcendental deduction of the maxims of reason (Ibid., pp. 549–50, B697–9), but also a transcendental deduction of the maxims of the reflecting capacity of judgement (KdU, pp. 22–4, 183–4 – II Introduction). Although I do not go into the detail of both deductions, I wish to focus on some aspects of the first one, since they reinforce the way in which we must intend the apparent contradiction between subjective and objective, and how it is important to proceed on the basis of *as-ifs*.

This transcendental deduction begins by emphasizing the difference between what is given to reason as an absolute object and what is given to reason as an object in the idea, that is, between assuming something absolutely ('*suppositio absoluta*') and only relatively ('*suppositio relativa*').[69] In the first case, concepts directly determine the object. In the second case, things in the world must be considered *as-if* they really were the way the idea indicates. This means that the idea is not constitutive, but it regulates our exploration on the constitution and on the connections of the object of experience 'in general'. Thus, the idea is not objective in the sense that it is in the objects; indeed from this point of view, it is subjective. Instead, the idea is objective since it refers to the fictive (*as-if*) object. This implies that the maxim brings us to a systematic unity, and allows us to extend knowledge. But, in this case, we must have the maxims of reason in order to proceed according to the ideas. 'This, indeed, is the transcendental deduction of all the ideas of speculative reason' (KdrV, p. 550, B699).

I move now to another extremely important point in the Appendix. Kant, after analysing, one by one, the three ideas of reason, and so showing their regulative use and their fictive status, makes an interesting remark related to the third idea 'which contains a merely relative supposition of a being that is the sole and sufficient cause of all cosmological series, [this] is the idea of *God*' (Ibid., p. 559, B713). This idea, due to the interest of reason, according to which we have to think *as-if* there were a sovereign reason which is the cause of any phenomenon and any series of phenomena, 'opens out to our reason, as applied in the field of experience, altogether new views as to how the things of the world may be connected according to teleological laws, and so enables it to arrive at their greatest systematic unity' (Ibid., p. 560, B714–5).

It should be noted that in this way 'the *purposive* unity of things' (Ibid.) is introduced as a consequence of the fact that we must think the world *as if* it were caused by a supreme being. And this is exactly the external purposiveness, analysed specifically in the *Critique of Judgement*. What is remarkable is that in this third *Critique*, Kant suggests something seemingly contrary. For, teleology and theology must be differentiated both in the sense of the non-derivability of purposiveness from the concept of God, and in the sense of non-derivability of the concept of God from purposiveness. We should, of course, avoid the pitfall of a *circulus in probando* (that is, in a diallelus or begging the question), arguing firstly purposiveness on the basis of the concept of God and then the concept of God on the basis of purposiveness (KdU, pp. 261, 381–2). However, by starting from the fictive idea of a supreme being, cause of everything, the idea of purposiveness can be introduced even if in a 'presumptuous' way (Ibid., p. 264, 383), so the same could be done in the opposite sense, that is, it can be claimed that:

> *The peculiar character of my cognitive powers* is such that the only way I can judge [how] those things are possible and produced is by conceiving, [to account] for this production, a cause that acts

according to intentions, and hence a being that produces [things] in a way analogous to the causality of an understanding.

(Ibid., pp. 280, 397–8)

Note that this not a *Beweis* of the existence of a supreme being, but only an indication of the fact that, due to the way in which our cognitive capacities are structured, we need to resort to a supreme being to have the concept of a world capturable by a system and, thus, to have the unity of nature (see Ibid., pp. 281, 399).

However, both in the Appendix and more systematically in the *Critique of Judgement*, Kant shows that referring to a final cause, or to a teleological relation, or to some '*nexus finalis*', must absolutely not be intended as a denial of efficient causality, that is, as a denial of the possibility of having a mechanical or physical relation, that is, a '*nexus effectivus*' (KdrV, p. 560, B715–16). This must be carefully considered, if one is not to misread both the Appendix and the *Critique of Judgement* as Kant's attempt to replace efficient causality with teleology. These are two ways of approaching empirical knowledge that, for Kant, must be equally taken into consideration. To put it another way, we can pose causal why-questions, but we can pose purposive why-questions as well, even if we are not obliged to look for both a causal and purposive answer in every case (on this point, see Boniolo, 2005).

2.2.7 The discovery of laws

We come now to one of the important questions: how can we formulate the empirical laws? Particularly in more recent times, some commentators, emphasizing a radical reading of a few passages from Kant, have claimed that he supports a certain strong (for example Buchdahl, 1971, pp. 26 and 33) or weak form (for example, see Butts, 1994, p. 277) of induction in the domain of discovery.[70] However, these are not particularly good interpretations of Kant's method of discovery, even if it is true that in certain passages he seems to be ambiguous. The whole architectonic of Kant's system and his whole critical philosophy, analysed from the point of view of the 'Copernican revolution' or from the point of view of the 'combination' (*Verbindung*) contained in the Transcendental Deduction, drastically excludes the possibility of an *intellectus ectypus* and, thus, of the inductivist reading. A concept, and a judgement, are not the result of an inductive abstraction from common features of what has been observed, but a product of the knowing subject's intellectual activity in rendering the world cognitively significant.

In order to further clarify this point and, thus, to arrive at the solution to the problem of formulation of laws, some propaedeutic steps are worth taking. First, I recall that: 'Judgment in general [*überhaupt*] is the ability to think the particular as contained in the universal [*das Besondere als enthalten unter dem Allgemeinen*]' (KdU, pp. 18, 179 – II Introduction).

In this passage, which surely can be thought of as one of the core points of the *Critique of Judgement*, Kant is claiming something that it is quite impossible to read inductivistically. He states that the capacity of judgement concerns thinking 'the particular as contained under the universal'. Note that this statement, according to which the particular always has to be seen in the light of a universal, admits two readings: a logical one and an epistemological one. For the logical reading, it says that, so that the systematic unity of knowledge can be reached, both the singular empirical judgements must be subsumed under the empirical laws, and the more particular empirical laws must be subsumed under the less particular empirical laws. For the epistemological reading, in any particular we cognitively grasp there is a universal that allows us to grasp it as that given particular. And this latter aspect is clearly stated by Kant in the First Introduction to the *Critique of Judgement*: '[...] insofar as this ability [to cognize nature] requires that we are able to judge the particular as contained under the universal and to subsume it under the concept of a nature' (Kant, *Erste Einleitung*, pp. 392, 202′–3′)

Let us return to the capacity of judgement, which is split into the 'determining capacity of judgement' and the 'reflecting capacity of judgement'. While the first capacity allows us to subsume the particular under a law already existing, the second one allows us to find the subsuming laws. We will consider the latter, which is particularly relevant to our discussion. If reason in its regulative use of transcendental ideas completes the task of the understanding, as far as the unity of nature and the systemacy of its laws are concerned, the reflecting capacity of judgement completes the understanding in another way. As we have said, the understanding, although regulating nature 'in general' through the three analogies of experience, leaves the particular empirical case indeterminate, that is, the regulation of the particular natures. At this point the reflecting capacity of judgement intervenes (KdU, pp. 36–7, 196 – II Introduction). But the reflecting capacity is not dealing with the object determined, as happens in the case of the determining capacity of judgement, but with an object particular. By reflecting on this object particular the universality determining it is found, that is, the particular empirical law making it an element of a given series is discovered (see also Ibid., pp. 290, 406).

Now, as Cassirer (1910, ch. 5) ironically comments, if one would like to call this induction, one is free to do so. Nevertheless, one should note that it is not the abstraction of the universal from the particular that is at issue, but the fact that by reflecting on the particular, one finds the universal which makes that particular knowable as that particular.

How – one reasonably wonders – can the reflecting capacity of judgement find the universal that subsumes that given particular, giving it cognitive significance? First of all, it cannot resort to something else since it must 'serve itself as a principle' (Ibid., pp. 266, 385), otherwise, it would be a determining capacity of judgement (Ibid., pp. 18, 179 – II Introduction). In other words, it must prescribe to itself those subjective principles, those maxims, that, then, allow it to reflect on the particular, and so to find the universal which makes it cognitively significant. This is its characterization; that is, this is its 'heautonomy' (Ibid., pp. 25, 186 – II Introduction).

Kant offers two maxims of the reflecting capacity of judgement:

> *The first maxim* [...]: All production of material things and their forms must be judged to be possible in terms of merely mechanical laws. *The second maxim* [...]: Some products of material nature cannot be judged to be possible in terms of merely mechanical laws. (Judging them requires a quite different causal law – viz., that of final cause).
>
> (Ibid., pp. 267, 287)

This means that the knowing subject, when facing a particular, reflects on it to find the universal which cognitively constitutes it *qua* that given particular. This is possible on the grounds of two guiding principles,[71] first by asking the question:

> (1) 'Which is the causal empirical law, of the type $\forall x Ax \overset{C^E}{\rightarrow} Bx$, which constitutes such a particular as a particular of a given causal series?' (*a causal why-question*).

And then asking the question:

> (2) 'Which is the purposive empirical law, of the type $\forall x Ax \overset{P^E}{\rightarrow} Bx$, which constitutes this particular as particular of a given purposive series?' (*a purposive why-question*).

Where $\overset{C^E}{\rightarrow}$ is a causal empirical implication and $\overset{P^E}{\rightarrow}$ a purposive empirical implication. Thus, first we reflect on the particular guided by the causal maxim. Then we reflect on the same particular guided by the purposive maxim (KdU, pp. 267–8, 387–8).

Briefly, both the causal and the purposive laws are found by reflecting on experience, that is, 'with' experience; they are surely not inferred inductively 'out of' experience. They are only hypotheses made regarding that possible universal contained in the particular in question: 'This I shall entitle the hypothetical employment of reason' (KdrV, p. 535, B675).[72]

It is also obvious that the hypothetical law so produced (that is, discovered, if you prefer) is contingent both regarding the laws of understanding, and regarding the first metaphysical principles. However, it is necessary since it is a law of nature, even if an empirical one.

Note also that its universality is different from the universality of a pure principle, concerning nature 'in general', or from the universality of the first metaphysical principles, regarding the possibility of mathematizing nature discussed by physics. It is a more limited universality; it concerns only a certain class of particular objects.

2.2.8 Causality and purposiveness

It could seem from the interpretation above that I am claiming that in the *Critique of Judgement* there is an equally balanced analysis of the causal and purposive aspects. This is exactly what I suggest, if the epistemological point of view is concerned, but it is not the case if the point of view is that of their applicability. If it is true that in the *Critique of Judgement* Kant assigns an equal epistemological value to the causal and purposive aspects, it is also true that here he pays more attention to the purposive aspects of nature, in particular to what he calls the 'natural purpose' (*Naturzweck* or *Zweck der Natur*), that is, to the organized natural being, that is, the physical organism not reducible to a machine:

> Hence an organized being is not a mere machine. For a machine has only *motive* force. But an organized being has within it *formative* force, and a formative force that this being imparts to the kinds of matter that lack it [thereby organizing them]. This force is therefore a formative force that propagates itself – a force that a mere ability [of one thing] to move [another] (that is, mechanism) cannot explain.
>
> (KdU, pp. 253, 374)

This is the internal side of purposiveness (*innere Zweckmässigkeit*), and it must be taken separately from the external side (*äussere Zweckmässigkeit*). The first concerns the object whose organization has to be seen as a function of itself. The second regards the object whose being and organization is considered as a function of something external. However, the internal purposiveness is one of the most innovative aspects of the third *Critique* in comparison with the Appendix, where only the external purposiveness is at issue.

Epistemologically, the causal aspect and the first maxim are as important as the purposive aspect and the second maxim. With reference to this point, since Kant speaks of the 'purposiveness of nature' (for example, Ibid, pp. 20, 181 – II Introduction), it would not be improper to speak of the 'causality of nature', of course without confusing it with causality of nature 'in general'. That this equal epistemological relevance is really claimed by Kant, is confirmed by numerous passages of the *Critique of Judgement.*[73] Of course, each of the two epistemological aspects has a different explanatory task. If I want to know the causal-why of something, I use the regulative principle of the *Kausalität der Natur*. If I want to know the purposive-why of something, I use the regulative principle of the *Zweckmässigkeit der Natur*.

However, each of these two principles, the two maxims, has limits to its application:

(1) it is impossible to think causality or purposiveness as contained in things, but they are my modalities to cognitively constitute them;
(2) I cannot approach the world only causally or only purposively (for example, Ibid., p. 267, 287);
(3) I do not know to what extent the causal approach could be successful (Ibid., p. 300, 415), even if I am sure that it could not be successful in determining the organization of the living beings (since the 'formative force' cannot be explained through the 'motive force', Ibid., p. 253, 374), but this cannot be proved (Ibid., p. 269, 388);
(4) I cannot, because of my limits as a human knowing subject, unify the two principles in only one principle (for example, Ibid., p. 287, 404).

To conclude, I, as a knowing subject, must always search for the two sides and, thus, I must try to apply both the causal and the purposive maxim (for example, KdrV, p. 560, B715–16; and KdU, pp. 236–7, 360).

2.2.9 Lawfulness

While analysing the second analogy of experience, it has been shown that it imposes a causal lawfulness on nature 'in general', but that it does not intervene regarding the particular natures. These are constituted as such by the empirical causal laws and by the empirical purposive laws, which have been found through the maxims of the reflecting capacity of judgement. For these maxims, by reflecting on the particular phenomenon, allow us to produce hypotheses that make that given particular cognitively significant *sub specie* instantiation of an empirical law. That is, they allow us the lawfulness of particular natures, by spurring us to produce the suitable cognitively constituting empirical laws

The principle of the second analogy does not tell us 'same causes same effects', rather 'every time that an alteration occurs, there is an irreversible and invariable temporal succession, of two spatial contiguous events, where one is cause and the other is effect'. We also know that the understanding does not specify the particular causal law:

> But apart from that formal temporal condition, [the second analogy of experience], objects of empirical cognition are still determined, or, if we confine ourselves to what we can judge *a priori* – determinable, in all sorts of additional ways. Therefore, specifically different natures, apart from what they have in common as belonging to nature as such [*Natur überhaupt*], can still be causes in an infinite diversity of additional ways; and each of these ways must (in accordance with the concept of cause as such) have its rule, a rule that is a law and hence carries necessity with it, even though the character and limits of our cognitive powers bar us altogether from seeing that necessity.
> (KdU, pp. 22–3, 183 – II Introduction)

We have just one nature 'in general', but we have infinite 'specifically different natures', which, according to the interest of reason in its regulative use, are to be unified in a systematic unity. Thus, nature at the empirical level is not made lawful by means of the pure laws of the understanding, but by means of the particular empirical laws found by reflecting through the maxims of the capacity of judgement. Lawfulness at the empirical level is made possible by the reflecting capacity of judgement which, analogously to the legislating understanding of nature 'in general', could well be considered the legislator of the specific natures.

I would like to repeat once more that the second analogy, or rather its principle, does not say anything like 'same causes same effects', that is, it does not state, or give us, the uniformity of nature, not even in the sense of nature 'in general'. On the contrary, the uniformity of nature is made possible at the empirical level, that is, at the only possible level where the explicit form of the particular causal laws can be found. However, be aware, the uniformity of nature, or what Kant calls the 'order of nature', is not the result of the finding of the empirical laws, but it is an implicit presupposition a priori of the two maxims guiding us to discover particular laws:

> Hence, though the understanding cannot determine anything *a priori* with regard to these [objects], still it must, in order to investigate these empirical so-called laws, lay on an *a priori* principle at the basis of all the reflection on nature: the principle that a cognizable order of nature in terms of these laws is possible [...] since without presupposing this harmony we would have no order of nature in terms of empirical laws, and hence nothing to guide us in using empirical laws so as to experience and investigate nature in its diversity.
>
> (Ibid., pp. 24–5, 185 – II Introduction)

Therefore, the reflecting capacity of judgement presupposes the uniformity of nature at the empirical level, since it is exactly here that it plays its role. It follows that nature, always at the empirical level, is made lawful through the causal and purposive laws which are found hypothetically by applying the two maxims.

2.2.10 Schlick's problem

Let us approach now the problem of the nomological validity of empirical universal judgements, that is, Schlick's problem: the problem of lawness. We must deal with a particular *a* and particular *b*. What will convince me that the universal statement of the type $\forall x Ax \rightarrow Bx$ with $a \in A$ and $b \in B$, is really a causal law of the type $\forall x Ax \overset{CE}{\rightarrow} Bx$, or a purposive law, of the type $\forall x Ax \overset{pE}{\rightarrow} Bx$? In other words, how will I know that I have a nomological universal and not an accidental universal? The answer is simple: the reflecting capacity of judgement! Its maxims spur me both to search for a (causal or purposive) universal subsuming under itself that particular, and to interpret it as a hypothetical law. As seen, we can formulate the wrong hypothetical law, both in the sense of having found the wrong nomological universal and in the sense of having mistaken an accidental universal for a nomological one. However, this is the price that the knowing subject must pay, since he has neither an *intellectus ectypus* nor an *intellectus archetypus*.

Note, however, that by taking into account the meaning of the maxims of the reflecting capacity of judgement, to consider nomological what is accidental is almost impossible. If I take a coin out of my pocket and see that it is a nickel, I would hardly claim that the statement 'All coins in my pocket are nickels' is an empirical law. This is a universal statement which, however, it is neither causal, nor purposive, while the maxims of the reflecting capacity of judgement spur me to search only for causal and purposive laws.

The case in which we have statements such as 'All ravens are black' is different. Is it accidental or nomological? Kant would have claimed without hesitation that it is a nomological statement, not because it establishes that being raven causes being black, but because being a raven in a given environmental circumstances implies being black, since there the natural purpose of being a raven implies being black.

In brief, nature 'in general' is lawful, because there is the legislating understanding a priori; the specific nature is lawful, because there is the legislating reflecting capacity of judgement, even if this can only produce hypothetical laws. Moreover, the pure principles of the understanding and the first metaphysical principles are nomological, because they are produced in such a way by the understanding. On the other hand, the empirical laws are hypothetically nomological because only in this way can the reflecting capacity of judgement succeed in using them in order to subsume the particular, that is, in order to give it cognitive significance, that is, to see it *sub specie* instantiation of an empirical law.[74]

2.3 The system of lawness and the constitution of lawfulness

Throughout this chapter we have followed step by step the rise and fall of philosophical expectations concerning the possibility of grasping what a law is.

It was claimed that a law would be an all-statement of the kind $\forall x Fx \rightarrow Gx$, but it has been shown that there are non-all-statements, especially in the biomedical domain, which difficulty cannot be recognized as laws. For example, either we refuse the status of law to the statement 'genes encode for proteins', or we must accept that non-all-statements can be laws, since not all genes encode for proteins.

It was claimed that a law would be an unrestricted universal, that is, that, on the one hand, it would not contain individual constants (that is, units of measure, values of universal constants, references to scientific samples, scientific proper names), and, on the other hand, it would satisfy Maxwell's requisite. Unfortunately, as we have seen, this is not necessary: we have statements containing individual constants, and statements infringing Maxwell's requisite. For example, 'All albino

Mus musculus are homozygote' contains an individual constant 'Albino *Mus musculus*', and is valid only for the space-time region in which the species *Mus musculus* lives. Therefore, either it is not a law, or we must accept that a law must not necessarily be an unrestricted universal.

It was claimed that a law would be a statement without a vacuously true antecedent. Bu we have a lot of statements that we consider as laws that have vacuously true antecedents, such as the inertia law. In this case too, either these statements are not laws, or we must accept that a law can have a vacuously true antecedent.

Naturally, not all the all-statements are laws; not all the unrestricted statements are laws; and not all the statements without vacuously true antecedents are laws.

The old regularists and the new regularists have not been able to find a good solution for all these possibilities. Therefore some of them have resorted to an epistemological axe: 'Laws can only be all-statements. Yes, we know that in the biomedical domain we cannot speak only of all-statements, but in the biomedical domain there are no laws, pace the biologists and the medical scientists'; 'Laws can only be unrestricted universals. Yes, we know that many scientific statements are not unrestricted, but they must not be considered as laws, even if they belong to well-accepted scientific theories'; 'Laws can be only statements without vacuously true antecedents. Yes, we know that some scientific statements have vacuously true antecedents, but they must not be considered as laws, even if they belong to well-accepted scientific theories.'[75] Some others, instead, have resorted to a still more drastic solution: 'There are no laws at all'. Are we really willing to use the axe, and cut out a lot of statements that are usually considered as laws but which we are unable to fit into our idea of lawness? Are we really willing to rule out the possibility of speaking about lawness? Are we really willing to accept the bizarre idea that since we are not able epistemologically to frame lawness we should simply eliminate it from our inquiries? Are we really willing to accept that all the Philistines (lawness) should die just because Samson (the old and new regularists) has decided to die (having realized that their approach is not good enough to solve the problem of lawness)?

It would seem then that the onus was unavoidably on the new metaphysicians. But the cost would be too high: we should have to admit the possibility of essences, or something equally mysterious. Are we really willing to accept baroque a priori arguments supporting metaphysical commitments on mysterious essences? It is as if we were willing to accept that, since we are not able scientifically to explain a physical phenomenon, we must resort to fate, or to a god, or to some other mysterious intrinsic necessity of the world. Should we, then, follow Nagel, and admit that the problem of laws is inevitably vague? Or, even worse, should we follow van Fraassen and conclude that there are no laws at all?

Nevertheless, as the history of western philosophy teaches us, there are more approaches than the Humean (the new and old regularist) and the pre-Humean (the metaphysical). There are many other possibilities, and there is nothing intellectually untoward in investigating them to see if they can offer something useful for our problem. In this light, I have analysed Kant's proposal and, as anyone can recognize, it is extremely powerful and promising. Precisely by taking this seriously I want to accept it as a starting point and to move on. But what does it mean to be Kantian with reference to the lawness problem? Only what Cassirer accepted in his 1910 *Substance and Function*:

(1) accepting Kant's 'Copernican revolution', according to which we know what we have the concepts of. As we have seen, this is an epistemological turning point, since the central role in knowledge processes is no longer played by the world to be known, but by the knowing subject and its conceptual apparatus which permits it to know the world by cognitively constituting it.
(2) accepting that what is important for lawness is no longer the analysis of a single statement, but the analysis of the system in which that statement is inserted.[76]

The *epistemological Copernican revolution* and the *methodological emphasis on the system*: these are the two pillars of Cassirer's Kantism developed in *Substance and Function*. And these are the two pillars of what I am going to propose in order to deal successfully with the issues of lawness and lawfulness. But let us proceed step by step.

To begin with, I must clarify some elements. It should be clear that the term 'nomologicity' is to be intended as synonymous with 'lawness', and I do not use it – as sometimes it has been misunderstood by some neopositivists and post-positivists – to refer to a deterministic law whose syntactical form is an all-statement. Instead, I will use it according to its original meaning,[77] that is, as referring to a law in general, deterministic or not, writable as an all-statement or not.

Secondly, in what follows my aim is not to find out under what conditions an all-statement such as $\forall x Fx \rightarrow Gx$ is a law, that is, the conditions for lawness or for nomologicity of a universal conditional. Instead, more generally, I will enquire into the conditions for the lawness of statements called *generics*, that is, statements such as '*F*s are *G*s'. These are statements that can be read from the point of view of both quantity and quality.

With reference to quantity, they can assert that all *F*s are *G*s ('Massive bodies are the source of the gravitational field'); or that some (also one) *F*s are *G*s ('Members of the class Aves fly'); or that all (or some) *F*s have the probability *p* of being *G*s ('Offspring both of whose parents have genotype Aa, have respectively 25%, 50%, and 25% of probability to have genotype AA, Aa, aa'). I know that not all the statements 'all *F*s are *G*s' are laws (for example, 'Cars have four wheels'); not all the statements 'some *F*s are *G*s' are laws (for example, 'Students in the Italian schools are French speakers'); not all the statements '*F*s have the probability *p* to be *G*s' are laws ('Students in an Italian college have a 36% probability to fall in love with a college-mate'). But the problem is exactly this: which of them are laws? How must a generic be characterized to be a law? What is the condition for its lawness?

With reference to quality, we have to add something else. To speak in terms of '*F*s are *G*s' can be too cryptic, since it is not clear what 'are' indicates, and this could lead us to misunderstand in what sense the *F*s are the *G*s. To avoid this risk, I suggest making the generic more explicit, and write

'*F*s are in the relation *R* with *G*s',

where *R* is the epistemological relation which can be (1) a linear proximal causal relation ('Charged particles are deviated from their rectilinear trajectories by an electro-magnetic field'); (2) a circular proximal causal relation ('The expression of the gene clusters encoding for a set of enzymes allowing the biosynthesis of the tryptophan are regulated by the tryptophan repressor which in turn regulates the gene expression'); (3) an ontogenetic causal relation ('Salivary glands are derived from the endoterm'); (4) a phylogenetic causal relation ('Members of family Hominidae are biped'); (5) a functional relation ('Ubiquitin marks proteins for degradation'); (6) a teleonomic relation ('Body plans are determined by the developmental control genes'); (7) a teleological connection ('Homo sapiens has sex to have offspring') and so on.[78] Therefore the problem is understanding which of the infinite lawlike statements '*F*s are in the relation *R* with *G*s' are really law. That is, what characterizes the lawness that turns some of them into laws. Let us begin.

Definition
By *lawlike statement* I mean any statement of the kind '*F*s are in the relation *R* with *G*s', that is, any generic statement that can be particularized with reference to:

(1) its quantity, and therefore making explicit whether it is an all-statement or a some-statement, a non-probabilistic or a probabilistic statement;
(2) its quality, and therefore making explicit the type of epistemological relation *R* linking the *F*s to the *G*s.

Epistemological problem (Schlick's problem, or lawness problem)
Which of the lawlike statements '*F*s are in the relation *R* with *G*s' can be considered as a law?

Definition
By *system of statements* I mean any set of statements linked by connecting rules that can be both formal (for example, classical logical rules), and informal (for example, rhetoric rules). Of course, any statement can be directly or indirectly (via other statements), formally or informally linked to any other. In the case that all the possible statements were directly or indirectly linked, we would have only one system: the $system_0$. We will consider only sub-systems of the $system_0$. Recalling that a semanticizing area is the set of rules sufficient to grasp the sense of a given concept, and that a rule is a statement, then the sub-systems of statements we are interested in are nothing but the semanticizing areas we met earlier. Among the possible semanticizing areas we consider those characterizable by the couple $S =< \{s_1, \ldots, s_n\}, C >$, where $s_1, \ldots, s_n$ are the n generic statements of the kind $s =$ '*F*s are in the relation *R* with *G*s', and *C* is the set of the connecting rules among the $s_1, \ldots, s_n$. It is worth noting that in a semanticizing area we have three levels of rules: (1) the epistemological rules linking concepts inside statements, that is, the relations *R* according to which the *F*s are *G*s; (2) the connecting rules linking the different statements *s* of the semanticizing area; (3) the statements themselves, for these are rules (in particular that ones synthesized in the given concepts).

By taking into account what has been said on the concepts as representation and as rule, we can easily arrive at the epistemological interpretation of the generic statements at issue.

Epistemological interpretation of the lawlike statements
'*F*s are in the relation *R* with *G*s' is to be intended in the sense that whenever I recognize, in a particular semanticizing area, something as

an *f*, since I have the concept *F* cognitively constituting that particular something as an *f* of *F*, I recognize it also as a *g*, since I have also both (a) the concept *G* cognitively constituting that particular something as a *g* of *G*, and (b) the rule allowing me to render cognitively significant the relation *R* between *F* and *G*.

Epistemological interpretation of the semanticizing area

Any semanticizing area is a representation of the part of the world it refers to. The offered representation must be understood especially in the sense that it allows us to render that part of the world cognitively significant. Analogous to the case of the representation offered by the concepts, the representation offered by the semanticizing area has both an intentional reference and an empirical reference. The intentional reference of a semanticizing area, which I call the *intentional area*, is the set of all the intentional objects and properties theoretically permitted by the latter. The empirical reference of a semanticizing area, which I call the *empirical area*, is the set of all the intentional objects and properties that are empirically detectable by whoever wants to detect them, that is, that are empirically objective. (Figure 2.1 gives a graphic representation of a system of statements.)

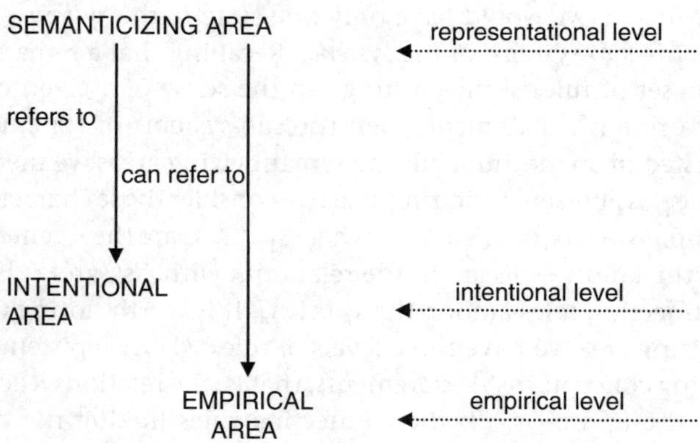

Figure 2.1 System of statements

Lawness

By *lawness* I mean the particular features of a lawlike statement that allows us to consider it as a scientific law. It is clear that lawness concerns neither single statement *qua* single statement, nor the representational level offered by the semanticizing area alone, nor the intentional area, nor the empirical area alone. Rather, lawness concerns the interplay between the representational level and the empirical level, of course mediated by the intentional level, that is, lawness concerns the interplay between the semanticizing area and the empirical area. Therefore a lawlike statement such that '*F*s are in the relation *R* with *G*s' is a law if:

(1) it belongs to a semanticizing area;
(2) the semanticizing area in question refers to an empirical area, that is, to a set of objects and properties that are empirically detectable by whoever wants to detect them, that is, that are empirically objective;
(3) the semanticizing area in question refers to an intentional area, some objects and properties of which indicate the possibility that there are also their empirical counterparts.

Point (1) focuses the role of the system; point (2) focuses the empirical confirmation of the semanticizing area; point (3) focuses the idea that a semanticizing area supports the validity of counterfactuals, that is, that it allows us the prediction of not yet known events. For example, the lawlike statement 'Incompressible fluids in stationary motion conserve their energy' is a law (Bernoulli's theorem), because it belongs to a semanticizing area (classical fluid-dynamics; or to a larger one: classical many-particle dynamics; or to a still larger one: classical mechanics; and so on) characterized by an enormous bulk of empirical confirmations, intended in the sense that there are lots of empirical objects and properties it refers to, and by a strong predictive power. On the other hand, the lawlike statement 'All mothers in the world are attractive' is not a law, since, even if it belongs to a semanticizing area, this is not characterized by empirical confirmation, in the sense that the properties it refers to are not empirically objective; moreover the system has no predictive power.

Remark on the empirical area

Of course, the intentional area is logically infinite, while the empirical area is finite. It is interesting to note that the logical infinity of the semanticizing area is usually pragmatically restricted by the knowing subject's needs. Moreover, the limits of the empirical area are delineated step by step through empirical inquiry, that is, through observations and experiments. In this way as well, *the validity area of the representation*, that is, of the semanticizing area, is going to be individuated. The validity area is not equivalent to the empirical area, since there could be the possibility of the existence of objects and properties, which we know from the knowledge of the intentional area, that nevertheless cannot be empirically detected due to their technological unobservability, as I argued in the past chapter. Nevertheless, step by step the validity area of a scientific representation is more and more individuated, and more and more it comes to be asymptotically close to the empirical area.

We know that observations and experiments can have both positive and negative results. The positive results concur for the empirical confirmation of the representation, while the negative ones concur for its empirical disconfirmation. However, both the positive and the negative results cooperate to delineate the empirical area and the validity area of the representation. At the intentional level we have both intentional objects and intentional properties, and we know that there could also be their empirical counterpart. To control this possibility, if it is technologically practicable, I must make observations and experiments, which, if I want their results to be empirically objective, must be repeatable by whoever wants to repeat them. At this point, a question arises: how can I understand if a negative empirical result is a *counterexample* or an *exception*?

To arrive at a good answer, we must reflect on the fact that the validity area of a representation is not only delineated, step by step, by observations and experiments, but also by the representation itself. For it is the representation itself that surely indicates what it does not represent. Do not forget that any representation tells us something about what is represented and something about what is not represented. Let us consider the statement 'Members of the class Aves fly'. We know that *Struthio camelus* (ostrich) and *Sphenicus magellanicus* (Magellan penguin) do not fly. These are not counterexamples of the statement above, but exceptions. We are aware of this not by analysing the statement itself but by knowing the representation in which that statement is inserted, that is, by knowing ornithology. From ornithology we know which members of the class Aves can fly, since to fly some structural and morphological properties must be satisfied. We also know which structural and morphological properties to be satisfied are from a different but connected representation, that is, aero-dynamics, which is a sub-representation of fluid-dynamics, which is, in turn a sub-representation of mechanics and so on.

It can happen that the ornithology we know does not indicate all the exceptions to the statement 'Members of the class Aves fly'. For it may happen that there are species of birds that we do not know yet and that do not fly. But this is an epistemic problem, and not an epistemological problem on what an exception is. At this point I can introduce two definitions.

Definitions

(1) By *exception* I mean a case that (a) does not satisfy a given law; (b) is contemplated as a negative case by the representation in which that law is contained.

(2) By *counterexample* I mean a case that (a) does not satisfy a given law; (b) is not contemplated as a negative case by the representation in which that law is inserted, and is found by means of observations and experiments.

Therefore we know whether a negative case is an exception or a counterexample not by considering the violated law alone, but by considering both the whole representation in which that law is contained and its negative detectability.[79] Moreover, both the exceptions and the counterexamples cooperate to delineate the validity area of a representation. But while the exception delineates the validity area from the theoretical point of view, the counterexample delineates it from the empirical point of view.

Lawfulness

By *lawfulness* I mean the nomological regularity of nature as a result of the cognitive constitution realized by the imposition of the semanticizing area. We know that according to the old and new regularists nature is not lawfulness, and only the conditions for lawness must be investigated, and that according to the new metaphysicians, nature is lawful due to its intrinsic essence. But I have shown the flaws in both these conceptions. According to my proposal, lawfulness of nature concerns the fact that nature has been rendered cognitively significant by semanticizing areas composed by laws. In a more Kantian terminology, nature is lawfulness in the sense of both *natura materialiter formata* and *natura formaliter spectata*.

Therefore, on the one hand, we must speak in terms of lawfulness, but, on the other hand, this does not regard the essence of nature (*there are no laws inside nature*), but rather that nature has been cognitively constituted in a certain way by means of the cognitive imposition of a given semanticizing area, composed of statements that, inside that semanticizing area, must be considered as laws. This is one of the most remarkable consequences of the acceptance of the Kantian 'Copernican revolution'.

Remark 1: On the organization of the semanticizing areas

Any particular representation offered by a system of statements $S = < \{s_1, \ldots, s_n\}, C >$, especially any given scientific representation offered by a scientific system of statements, allows us a specific way of organizing the compound statements by means of the connecting rules *C*. We know that a different system of statements $S' = < \{s'_1, \ldots, s'_n\}, C' >$ allows a different way of representing the world and of organizing the compound statements, now via the connecting rules C'.

There are certain cases in which we can change the representation by changing the connecting rules alone, that is, moving from *C* to C', even if it would be extremely difficult to maintain that we have changed the system of the statements as a whole. In this cases, nevertheless, it can happen that statements before considered to belong to the system *S*, since they were connected by means of *C* to other statements, are no longer considered to belong to the system, since now – that is, from the point of view of the connecting rules C' – they are no longer connected to other statements of the systems.

A typical example is offered by special relativity. If we compare the system of statements contained in Einstein (1905) and in Landau and Lifšitz (1967), we note a non-trivial difference. Certainly, special relativity is described in both cases, but in Einstein's seminal paper there were some rather vague notions, in particular those concerning the transverse mass, the longitudinal mass, and, of course, the rest mass. Instead, the way of presenting special relativity via four-dimensionality and variational principles offered in Landau and Lifšitz's text allows us to understand that the transverse and the longitudinal masses are vestiges of a pre-relativistic approach to electromagnetic phenomena. They are no longer present. What has changed from the former case to the latter is, first of all, the set of connecting rules.

Remark 2: On similarity and isomorphism

We know that a representation to be a representation must represent something. This simple and trivial fact is contained in the definition itself of the term 'representation'. Therefore there should be a relation between the representation, that is, what represents, and what is represented. But what is the status and what are the features of this representation relation? It depends, of course. In particular it depends on the particular representation we are dealing with, and on the particular epistemological background we share.

In our case, since we have embraced a Kantian point of view, the representation does not mirror the world itself, or a part of the world itself. What is represented is not a part of the world itself, but the part of the world constituted in a cognitively significant way by the representation. What is represented is, as seen, both *natura materialiter spectata*, that is, the set of all the phenomena, and *natura formaliter spectata*, that is, the set of the nomological relations among phenomena. This means, if it is necessary to repeat it again, that in my approach the representation stands, first of all, in a relation of cognitive constitution with what is represented.

Needless to say this claim does not imply that what is represented necessarily exists. There are two levels of what is represented: the intentional level and the empirical level. Moreover, since a representation represents something beyond itself, there are some similarities between the representation and what is represented both in the intentional area and in the empirical area. Of course there are some kinds of morphisms (intended in an algebraic sense) between the representation and the two represented areas. But these similarities and morphisms must be considered cum *granum salis*, that is, there is neither a strict isomorphism nor a mirror-like relation:[80] we know very well that the human knowing subject has no *intellectus ectypus*. However, what is more important is that we work theoretically inside the representation by drawing inferences, logically reorganizing, enlarging some aspects, cutting some other aspects away and so on. Therefore any representation is the matter on which and by which an agent (a scientist) theoretically and empirically works better to represent the part of the world at issue. Any representation is both *id quo conoscitur* and *id quod conoscitur*.

3 Theories, Models, Thought Experiments and Counterfactuals

Discussing scientific theories and scientific models could be seen as old hat; something on which nothing new can be said. However each different interpretation is a different point of view, and each point of view enables us to see completely new aspects, or old aspects in a new light.

In this chapter I will propose an interpretation of this supposed old hat, in the hope that it will indeed highlight new aspects. In particular I will suggest considering scientific theories, following an ancient and well radicated tradition, as *hypothetical representations*, that is, as conceptual tools designed for cognitive aims; and models as fictions, or, rather, as *fictive representations*, that is, as as-if constructions designed for pragmatic aims. This will be argued by reference to some seminal works from the dawn of the contemporary philosophy of science and will be followed by a suggestion for a taxonomy of the models.

After theories and models, I will move to thought experiments, which I will discuss on the basis of the idea of fictions. These will be divided into two classes: (a) the *rhetorical thought experiments*, an argument starting from an as-if world and arriving at certain conclusions which cannot be empirically tested; and (b) the *exploratory/clarifying thought experiments*, which are used either to explore regions of a theoretical domain that is not completely clear, or to clarify the meaning of a certain concept or a set of concepts.

Underlying the ideas of both fiction and the as-if world there is, of course, the notion of the counterfactual. The third part of this chapter will be devoted to this. But it will not be approached using the usual logico-linguistic tools. Rather, a hermeneutical approach, derived from the work of Weber and Vaihinger between the end of the nineteenth century and the beginning of the twentieth, will be proposed.

3.1 Theories and models: old hat?

3.1.1 Theories as hypothetical representations of the world

The interpretative proposal I am suggesting finds its roots in the pre-World War II European philosophy of science, in particular in that presented by two physicist-philosophers at the end of the nineteenth century, Hertz and Boltzmann. For them, scientific theories are, first of all, to be thought of as *Bilder*, representations, that is, something, objectively *built* (*gebildet*) by the knowing subject according to certain rules, allowing the subject to intellectually orient itself in the world, or, rather, in a part of the world.

In the great introduction to *The Principles of Mechanics*, Hertz, again taking up Kant's 'Copernican revolution', develops an interpretation of scientific theories according to which they should be considered neither as offering a true description of the world, nor indeed as the most truth-like one, but – as Boltzmann will claim later – 'as a mere picture [that is, representation – *Bild*][1] of nature [...] which at the present moment allows one to give the most uniform and comprehensive account of the totality of phenomena' (Boltzmann, 1899, p. 83).[2] First of all, this means considering theories not as true or truth-like formalizations, but as attempts made by the knowing subject to grasp the empirical data. It follows that we should be looking neither for the essential structure of the world, nor to moving closer and closer to such an essential structure. Instead, we are the constructors of representations which allow us to grasp empirical facts, to give them cognitive significance. However, according to Hertz and Boltzmann, the role of representations is not limited to recording in a coherent and economic way what is already empirically known. Representations also have to predict future events (remember Schlick's criterion for lawness).

Therefore, it is not surprising that Hertz opens his introduction with an emphasis of this point:

> To draw inferences as to the future from the past, we adopt the following process. We form for ourselves images [representations] or symbols of external objects; and the form which we give them is such that the necessary consequents of the images [representations] in thought are always the images [representations] of the necessary consequents in nature of the things pictured. In order that this requirement may be satisfied, there must be a certain conformity between nature and our thought.
>
> (Hertz, 1894, p. 1)

Thus, the representations do not tell us how things are in themselves, but, Kantianly, how 'our conceptions of things' are (Ibid.). Representations are both what I know, and that by means of which I know: again *id quod conoscitur* and *id quo conoscitur*.

Someone might object that if the representations are our constructions it is not clear how there can be strong relations between them and the empirical world. To answer, it is necessary to bear in mind the three criteria that, according to Hertz, allow us to choose the best representation (*but not the true or the most truth-like one*) among those we can construct.[3]

The first criterion, the *requirement of permissibility*, states that only those theories which do not 'contradict the laws of thought' must be considered. That is, there must be coherence between the representations and the laws of thought.[4]

Then, among the permissible theories, those which are not contradicted by experience must be chosen, that is, those satisfying *the requirement of empirical correctness*. Moreover Hertz is perfectly aware that scientific representations are hypothetical constructions made to grasp reality and, therefore, that reality plays the main role in evaluating them. Only by appealing to experience can they be confirmed or refuted: '[...the representation] has the character of hypothesis which is accepted tentatively and awaits sudden refutation by a single example or gradual confirmation by a large number of examples' (Ibid., p. 36).[5]

Finally, among the theories that are both permissible and empirically correct we must address our attention towards the one that satisfies *the requirement of appropriateness*. That is, (a) the most distinct one, in the sense that it includes the greatest number of essential relations, and (b) the simplest, in the sense that, besides the essential relations, it contains the smallest number of superfluous and empty relations, even if 'empty relations cannot be altogether avoided: they enter into the images [representations] because they are simply images [representations], – images [representations] produced by our mind and necessarily affected by the characteristics of its mode of portrayal' (Ibid., p. 2).

Undoubtedly, the third requirement seems both ambiguous and vague, as Hertz himself writes. Before determining the most appropriate representation we should determine what is really essential (as far as the distinctness is concerned), and what is really less divergent from such an essentiality (as far as the simplicity is concerned). However, the other two requirements, especially the first, are not trivially interpretable either, and therefore are not easily applicable to settle the underdetermination. However, let us leave aside this interpretative question and content ourselves with the fact that we have found two intellectual ancestors in Hertz and Boltzmann who support the idea that the theories are representations (although neither true nor truth-like) constructed both to hypothetically represent the phenomena in a comprehensive and homogeneous way and to hypothetically predict new ones. It should be remarked that this approach[6] provides us with a good *antirealism on the theories* (but *not an antirealism on the entities*), teaching us that *a scientific theory is neither a mirror copy of reality nor its approximation but a hypothetical representation constructed by the knowing subject to know and live.*

3.1.2 Models as fictive representations of the world

Let us move to the models, which I argue to be fictive representations. It should be noted immediately that, according to some authors, interpreting scientific theories as representations means interpreting them as models. Already this remark should make us aware that the term 'model' is not always used in the same way. First of all, a model can be something whose structural and/or functional relations are something to aspire to, that is, it may be thought of as an ideal-type à la Weber. I will come to that later. Second, it may be something whose structural and/or functional relations are reproduced in some way by other structural and/or functional relations (for example, the model of a statue is what the statue reproduces). Finally, a model may be something whose structural and/or functional relations allow us to grasp and organize known empirical data, and to predict new ones, although it is not necessarily reproducing in some way the structural and/or functional relations of the elements of the world. It is this third sense that interests us.

However let us return to the idea that scientific theory itself may be interpreted as a model.[7] In this way misunderstandings may arise. Boltzmann, in the entry on 'model' in the *Encyclopaedia Britannica*, explains: 'purely abstract conceptions should be helped by objective and comprehensive models in cases where the mass of matter cannot be adequately dealt with directly' (Boltzmann, 1902, p. 215). This means that to acquire knowledge we need, beyond the theories – which are general and abstract representations – particular representations which allow us to gain a better understanding of a given set of empirical facts. By taking into account this aspect and for the sake of clarity, we should distinguish the level of general abstract representations (theories) from the level of particular representations (models, in Boltzmann's sense). It follows that if we wish to continue interpreting the theory-representation as a model, we should take care to emphasize that it is a model of models, that is, a large representation containing smaller and sometimes independent representations.

Therefore, to avoid possible misunderstandings arising from these different uses, I suggest not using the term 'model' to refer to the theory, and emphasizing that the representations offered by the theories are epistemologically different from the representations offered by the models. The former are *hypothetical representations*, the latter are *fictive representations*.

3.1.3 As if the world were

Speaking about fictive representations means coming back to the approach proposed in 1911 by Vaihinger in *The Philosophy of 'As If'*, which had (and has) the sad fate of being more quoted than read. Many authors affirm that the German philosopher claims that 'all is fiction'. Unfortunately this is not true. Actually, Vaihinger is far subtler than this, and according to him not all is fiction, and not all fictions are similar.

His theses are more complex than this would lead us to believe, and his fictionalism is a direct consequence of his historical studies. He was a leading character of the *Kantphilologie* and, between 1881 and 1892, author of a huge *Commentar zu Kants Kritik der reinen Vernunft*. The source of his fictionalism springs from the pages of the *Critique of Pure Reason* in which Kant states that the ideas of pure reason – that is, the ideas of soul, world and God – are 'as if' having the pragmatic (and non-theoretical) aim of leading research.[8]

> *Fictio* means, in the first place, an *activity* of *fingere*, that is to say, of constructing, forming, giving shape, abstracting, presenting artistically, fashioning: conceiving, thinking, imaging, assuming, planning, devising, inventing. Secondly, it refers to the *product* of these activities, the fictive assumption, fabrication, creation, the imaginal case.
> (Vaihinger, 1911, p. 81)

In this way Vaihinger explains the meaning of the term 'fiction' that is at the core of his theory of knowledge.

According to him, to gain knowledge, the human being constructs conceptual representations of the world which are totally different from the actual world. They do not mirror the actual world at all. Moreover there are three classes of conceptual representations: *dogmas, hypotheses* and *fictions*. Only the third of these have a relevant pragmatic role, satisfying the aim for which they have been constructed, since they allow us to orient ourselves in the world and, thus, to predict future events. As fictions have only pragmatic aims, they should not be considered as theoretical concepts, but, rather, as pragmatic concepts.

There is a theoretical and an epistemological gap between the actual world in which we want to orient ourselves, also predictively, and the fictions we construct to realize it. However, any such gaps vanish when the actual world is pragmatically replaced with the imaginary world represented by the fictions. Even if the fictions represent an as-if world, they are a (secondary) product of the actual world in which they allow us to live: 'The representations are not at all pictures [in the sense of mirror copies] of the becoming; yet they themselves are the becoming; rather, a part of the cosmic becoming' (Ibid., ch. 13). Thus, the human being and human thought belong to the actual world whose imaginary representation is given by the fictions. Unavoidably, it follows that the aim of the thought is not the thought itself but its fictive products, which facilitate action and life.

Even if the fictions are mere imaginary constructs different from the actual world, that is, pure as-if constructs, it should not be surprising, from an evolutionary point of view, that they satisfy their pragmatic aims so well. For throughout its long history, humanity has been constructing fictions which have increasingly satisfied the necessity to act and live. That is, if the core of fictionalism is given by the Kantian representationalism, totally freed from apodicticy and transported to pragmatic shores by means of the emphasis on the status of transcendental ideas, Vaihinger, in order to justify the pragmatic success of the imaginary concepts, has recourse to a typical evolutionary approach, even if coloured in a teleological way.

It is necessary to underline that, according to Vaihinger, fictions, and therefore fictive judgements, are completely different from hypotheses, and therefore from hypothetical judgements. Hypotheses are conceptual representations which, *conjecturally*, tell us *how the world is*, while fictions are conceptual imaginary representations by means of which *we postulate an as-if world*. Therefore, the aim of hypotheses is epistemologically different from the aim of fictions. Being attempts to representatively grasp the actual *world without exceptions*, hypotheses have a cognitive and theoretical aim, while, as stated above, fictions have pragmatic aims, being instruments to allow us to act and live in the actual world, of which they are *an exceptional representation*. This means that while hypotheses can be refuted when contradicted by the world they try to grasp, fictions cannot, being contradicted by the actual world from the outset. The latter are constructed by the knowing subject on false premises.

Therefore, both of them are representations, but the hypotheses are not intentionally constructed against the facts, while the fictions are intentionally constructed against the facts. However, we can also deduce consequences from fictions. If these do not match the world, the fictions must be rejected. Yet this rejection is not made for theoretical and cognitive reasons (as happens in the case of hypotheses), but for pragmatic reasons: *an instrument is simply set aside when it is no longer useful for the aim for which it has been constructed* (Ibid., ch. XXI), and fictions are instruments. Analogously, while we are theoretically and cognitively justified in maintaining a hypothesis confirmed by data, the only justification in maintaining a fiction is given by the pragmatic 'utility it gives to the empirical science' (Ibid.).

3.1.4 Theories and models: two kinds of representation

At this point, I can summarize what has been said about hypothetical representations (theories) and fictive representations (models).

Prima facie, scientific theories are hypothetical representations allowing us (1) to organize in a general, abstract, and formal way the empirical facts we already know; (2) to predict, at least statistically, new empirical facts; (3) to give cognitive significance to old and new empirical facts.

Nevertheless, if we really want to arrive at the empirical facts we must focus such general and abstract theories onto the specific empirical situation we want to analyse. This particularization is facilitated by a particular kind of model, the *focusing model,* by means of which we consider only those aspects of the analysed situation that we conjecture to be necessary to deal with it successfully. That is, we construct a fictive representation by particularizing the mother-theory, that is, the hypothetical representation. It is exactly by means of such a fictive representation of an as-if world that we link the hypothetical representations with the empirical facts.[9]

However, it may be the case that the problem to be solved is too complex to tackle from the starting point of a hypothetical representation, or it may seem impossible for thought to succeed in disentangling the facts under analysis. In such situations, we substitute the hypothetical representation with a fictive representation, making the problem tractable. Thus we use another kind of model again: the *replacing model.*

A third type of model enters the scene when there are facts which cannot be arranged according to any existing hypothetical representation, and, at the same time, we are unable to propose a new one. To organize these, we must construct a fictive phenomenological representation, that is, a *phenomenological model*.

Finally, whenever we deal with objects, we are dealing not with real objects but with idealized, and therefore fictive, objects, that is, with *model objects*.

In all these cases, fictions are not tentatively proposed as theoretical tools to grasp the actual world, but as pragmatic tools to solve specific problems and/or to permit the link between the general and abstract level of theories and the level of the empirical facts. All of this can be schematized as in Figure 3.1.

Before discussing and exemplifying these possibilities, two points are worth emphasizing. First, whenever I affirm that a theory represents hypothetically *how the world is*, while a model represents an *as-if world*, implicitly I claim that there is some similarity between what is represented, respectively, by the theory and the model and the world. Yet, to reiterate, this is not similarity as in a mirror image or as isomorphous similarity, but a constitutive similarity, in the sense that the constructed hypothetical or fictive representations are similar to what is cognitively constituted in that way in the world. However,

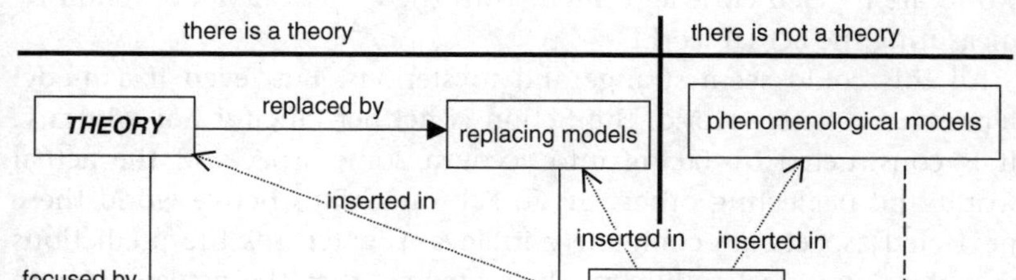

Figure 3.1 Relations between theories, models and reality

the constitutive similarity is epistemologically different in the two situations. As Vaihinger asserts (Ibid., chs. 4 and 22), in the first case we have a *hypothetical similarity* between the hypothetical representation and the world *qua* cognitively constituted world; in the second case we have a *fictive similarity* between the as-if world and the world *qua* cognitively constituted world.

Second, working with models can be surprising, because from a fiction that we know to be different from the world and which is also in some part constructed against the facts of the world, as Vaihinger and many others have observed (for example, Turing 1952), it is possible to infer conclusions which are in good agreement with the actual world. An emblematic example is given by the model of the universe proposed, within general relativity, by Schwarzchild in 1916. The fictive world represented by this model is very different from the usual conception of the world around us. It comprises a world composed of a non-rotating spherical body and nothing else. Nevertheless, in this model things work so well that the correct values for the three classical tests of general relativity are obtained (the gravitational red shift, the deflection of the trajectory of a photon approaching the source of a gravitational field, the perihelion precession of Mercury). That is, when the empirical tests were carried out it was found that *the empirical values are 'more or less' the ones found as consequences of Schwarzchild's fictive world*. How is it possible that predictions deduced from a model representing a completely fictive world are in such close agreement with the empirical values found by measuring the actual world?

All this could seem strange and mysterious. But, even if a model represents a fictive world, the fiction is neither fanciful nor arbitrary. It is constructed by taking into account some aspects of the actual world and neglecting others. As in Schwarzchild's fictive world these neglected aspects are completely irrelevant as regards the predictions and their empirical fulfilment. It is obvious that the actual universe is not made up of only a non-rotating spherical body, but adding other elements would be entirely irrelevant as to the problem at issue.

This introduces a necessary heuristic principle which allows us to construct models:

> *the principle of the fictive 'abstrahere ab aliquo'*,[10] according to which *the knowing subject, while constructing models, cuts away from the world all those aspects he/she thinks irrelevant to solve that specific pragmatic problem.*

It is a necessary principle, because without it models cannot be constructed. It is a heuristic principle, because it leads towards their construction. It is fictive, because the knowing subject constructs an as-if world in full awareness that the real world is not so.

This is a principle that, very probably, was formulated explicitly for the first time by Galileo Galilei (1632, pp. 208–9):

> SALVIATI. Then whenever you (Simplicio) apply a material sphere to a material plane in the concrete, you apply a sphere which is not perfect to a plane which is not perfect, and you say that these do touch each other in one point. But I tell you that even in abstract, an immaterial sphere which is not a perfect sphere can touch an immaterial plane which is not perfectly flat in not one point, but over a part of its surface, so that what happens in the concrete up to this point happens the same way in the abstract. It would be novel indeed if computations and ratios made in abstract numbers should not thereafter correspond to concrete gold and silver coins and merchandise. Do you know what does happen, Simplicio? Just as the computer who wants his calculations to deal with sugar, silk, and wood must discount the boxes, bales and other packings, so the mathematical scientist (*filosofo geometra*), when he wants to recognize in the concrete the effects he has proven in the abstract, must deduce the material hindrances, and if he is able to do so, I assure you that the things are in no less agreement than arithmetical computations. The error, then, lies not in abstractness or concreteness, not in geometry or physics, but in the calculator who does not know how to make a true accounting. Hence if you had a perfect sphere and a perfect plane, even though they were material, you would have no doubt that they touched in one point; and if it is impossible to have these, then it was quite beside the purpose to say *sphaera aenea non tangit in puncto*. But I have something else to add, Simplicio: Granted that a perfect material sphere cannot be given, nor a perfect plane, do you believe it would be possible to have two material bodies with their surface curved in some places as irregularly as you pleased?

Yes – Salviati-Galilei says – we know that real spherical objects touch real tables in many points, but we must abstract from this situation (in the sense of *abstrahere ab aliquo*), and consider models of objects: a perfectly spherical ball and a perfectly plain table. In this case there is only one contact point. We must cut away the superfluous; we must 'eliminate the hindrances', and in this way we arrive at a model which is a fictive abstraction. Note that Galilei was speaking about a model object, but the same consideration is valid for any kind of model, and on this point Poincaré (1902) would have peremptorily claimed, *'Les savants n'ont jamais méconnu cette vérite'* – scientists never underestimated this truth. A position that was widely accepted in the pre-World War II European philosophy of science, as this quotation from the great Italian mathematician Volterra (1901–02, p. 10) affirms:

> To study the law of the variation of the objects subject to measurement, to idealize them by depriving them of certain properties [that is, by abstracting from their non-interesting properties, or *abstrahere ab aliquo*] or by attributing to them other properties in an absolute way [that is, by abstracting their interesting properties, or *abstrahere aliquid*] and to establish one or more elementary hypotheses ruling their simple or complex variation: this marks the initial moment at which the real foundations, on which the entire analytic building can be constructed, are posed. It is exactly now that the entire power of the methods that mathematics renders available to those who are able to use them glow.

Certainly, there is a gap between the as-if representation offered by a model and how we, also intuitively, think that that part of the world fictionally represented by the model is. But if the model is a good model, this gap is totally absorbed by the *error interval*. This is important: *the pragmatic validity of a model must be discussed considering its empirical confidence, that is, to what extent the numerical values we find by working with it match the empirical values we actually find by means of the measurement apparatuses*. But the empirical values are obtained within an error interval which is not only due to the imperfection of the measurement apparatus, but also to the perturbations caused by the rest of the actual world, that is, the part of the world not considered in the model. Yet usually these perturbations are irrelevant as to the problem with which the model is concerned. The distance between the as-if world represented by the model and the real world fictionally grasped is not important if the former is able, within the error interval, to organize and predict the empirical values obtained or obtainable in the latter.

The initial surprise should have vanished now. The conclusions derived from a model, that is, from a fiction, are in close agreement with the empirical observations because, and as long as, the error interval of the measurements absorbs the effects due to aspects of the actual world neglected in the model.

A taxonomy of the models

Following what we have said above, we may divide the models into three main classes: (1) *principle models*, (2) *model objects*, and (3) *phenomenological models*. I also wish to consider *ideal-typical models*.

Principle models

By *principle models* I mean those models which, owing to the maintenance of the characteristic structure of theories (that is, principles and consequences), seem to offer a hypothetical representation equal to the one offered by the theories. Actually, *they offer only a simulacrum account of the hypothetical representation*: being models, the world they represent is not the real world, but an as-if world which is fictionally, and therefore pragmatically, considered instead of the actual world.

These models are the best known, but it is worth saying something about them. First of all, it seems that they belong in particular to physical sciences, where, unlike in other scientific domains such as the biomedical or chemical domains, there are well-structured (from a logical point of view) theories. Second, they may be sub-divided into (1) models focusing a theory otherwise not able to grasp particular empirical situations (*focusing models*); (2) models replacing an entire theory whenever it is impossible to apply the latter owing to the complexity of calculus it involves (*replacing models*).

Focusing models. In this case, we are dealing with models which allow us to grasp a particular empirical situation by focusing the general and abstract theory-representation from which they are deduced. At the same time, these models enable the link between such a theory and the empirical level. Of course they do not contradict the hypothetical representation from which they are deduced. Their premises are directly derived from the premises of the correlated hypothetical representation by imposing certain constraints which, on the one hand, focus them on the issue in question, and, on the other hand, idealize the situation. Therefore, although they are, in this sense, coherent with their mother-theory, they diverge from the real world. It is the process of focusing and idealization that leads to the construction of a fiction.

Focusing models may be exemplified in an emblematic way by general relativity. The mathematical structure of general relativity is a differential manifold with a Riemannian locally pseudo-Euclidean metric. Assigning a metric to the space-time structure means assigning peculiar relations among its points. Therefore, studying the explicit form of the metric tensor is equivalent to studying the gravitational field. Such an explicit form is obtained by solving the field equations, which are differential equations univocally solvable, by means of the Cauchy theorem, only by starting from well-determined initial conditions. It means that there are as many different solutions as different initial conditions, and each particular solution is connected with a particular model of universe.

Thus, on the one hand, there is the general and abstract representation of general relativity, culminating in the field equations, but unable to represent specific universes. On the other hand, there are the focusing models, deduced from it, which are able to satisfy such a specific request. Therefore, each model derived from this representation necessarily represents an *as-if universe*, as in the Schwarzchild model we have already seen.

Replacing models. Replacing models are constructed parallel to and analogical with the theories they replace. Therefore, they maintain unaltered the logical structure of the latter. However, representing a fictive world, they offer a simulacrum account of a representation of the actual world. That is, it seems that they represent the real world, but actually they deal with an as-if world that fictionally substituted the world represented by the replaced theory. These models are used when it would be too complex to handle the theory in a satisfactory way. Thus their premises go against the facts, just because they are models, but they also go against the premises of the replaced theory. This is the main distinction of focusing models.

The kinetic theory of gas gives a classical example of replacing models. We could have a physical representation describing a system of gas, but this would not be easily tractable for predictive aims. Thus, we devise a fiction considering the gas molecules as rigid spheres, whose mean distance is greater than their diameters, and whose elastic collisions are governed by the laws of classical mechanics. At this stage, the fiction is not very good, because the situation is still non-tractable.[11] To overcome this obstacle we must add, as Maxwell did, three statistical assumptions to the original fiction, that is, three new fictions: (1) the distribution of particles is uniform; (2) each component of the velocity is independent of the other ones; (3) the velocity probability distribution is isotropic. By means of these three assumptions, it becomes possible to construct an easily tractable model that can predict physical situations.

At the beginning, we had a perfectly deterministic physical situation which could be represented in a perfectly deterministic way by a representation. But this representation would have been non-tractable owing to the enormous number of variables in play. Therefore, we have been forced to construct a mathematized model representing a fictive physical world, an as-if world. Usually, a replacing model is also particularized by means of the focusing models, which enable the link between the level of the general and abstract assumptions of the replacing model and the level of the empirical facts.

Model objects

Model objects are atypical in the sense that, unlike the principle models and the phenomenological models that I am going to discuss in the next section, they do not have empirically testable consequences. Instead, they enable the theories or the models into which they are inserted to have empirically checkable consequences. They are no more than fictive constructions by means of which we idealize real objects (recall Galilei's sphere and table). By means of them, *we make regular* (that is, tractable by *regulae*) *the non-regular real objects of the world*. In other words, practising science also means, as Galilei taught us, to 'deduce the material hindrances', that is, replacing the actual 'dirty' objects with clean and more easily (sometimes more mathematically) handled fictions such as the rigid body, the point mass or the point charge, the spherical distribution, the ideal population and so on.

It does not matter if these objects are idealized and fictive. At this point, what I have said about the match between the consequences of a model (that is, the consequences of a fiction) and the empirical facts should be recalled. There is nothing strange in the fact that science, in spite of dealing with idealized objects, is able to successfully represent actual objects. Again, what enables such success is the play of the experimental errors, which are necessarily connected with each measurement and observation. We can use a *particular* model object (for example, the fiction of the massive point, or of the spherical mass distribution) of that *particular* actual object (for example, the Sun) since in that *particular* problem the divergence between fiction and reality is totally absorbed by the empirical errors. *The secret of the success of science lies exactly in the error interval linked to the experiments and the observations.*

I wish to exemplify the model objects by discussing the case of the continuous charge distribution, because it allows me to underline the limits within which they can be applied.

Using the continuous charge distribution means introducing a scalar function: the charge density $\sigma(x, y, z)$. It is obvious that this is a fiction: the source of the electric field is not a sort of charged magma but a system of charges with finite, even if extremely small, dimensions. The convenience of using this fiction is evident when we have to tackle great systems, but it becomes clearer when we realize that, by means of this model object, one is able to solve the problems connected with the presence of certain mathematical singularities. For near an actual charge the field $E = \sum_j (q_j r_j)/r^2 \rightarrow \infty$ as $(1/r^2)$. This obstacle can be overcome either by considering that it is physically non-significant to calculate the field inside the charge, or by adopting the fiction of the continuous distribution. In this case, the field $E = \int (\sigma(x', y', z')/r^2) dx' dy' dz'$ no longer goes to infinity, not even inside the distribution, where the volume element goes as $r^2 dr$, for $r \rightarrow 0$. However, this fiction, like the other fictive objects, can be conveniently used only in certain situations. For example, when the system is thought of *as if* it were sufficiently great to consider the element of volume $dxdydz$ sufficiently small, but containing a suitable number of charges such that the charge of the volume is $\sigma dxdydz$. Moreover, the fiction also allows for some conceptual adjustment. For example, the Poisson equation ($\nabla^2 \Phi = -4\pi\sigma$) is a differential equation that should be valid only locally, being the relation between the charge density in a point and the potential in an infinitesimal interval around this point. But we know that the continuous charge distribution is a fictive object. This fact leads us to interpret only macroscopically the Poisson equation and to consider the charge distribution as an average value in a small (but sufficiently large to contain many elementary charges) region.

In this case, we have a fictive object that can be used in certain phenomenological or theoretical situations. But we also have the case in which a phenomenological model works only for certain fictive objects, as happens for fictive populations in the case of models constructed in population genetics.

Phenomenological models

While principle models maintain the typical structure of a theory, so that they may be thought of as a simulacrum account of a hypothetical representation of the world, phenomenological models do not have any such structure. They might be considered more as a sort of spider's web constructed in order to capture observable facts that would otherwise be without any correlation. In this case, the scientist has neither a hypothetical representation for grasping the facts, nor a principle model capable of organizing them. Therefore, he or she is forced to create a model, practically from nothing. The scientist borrows mathematical and non-mathematical pieces from here and there trying to put them together in order to construct something which will allow correlation of the data. Thus, the scientist constructs a fiction, that is, a phenomenological model, hoping to be capable of *saving the phenomena*.

Thus, I am suggesting that the only real case in which a scientist *saves the phenomena* occurs when working with phenomenological models.[12] These have no purpose beyond that of adequately patching a theoretical incapacity to find deeper explanations by means of more fundamental (only in a methodological sense) laws. They are fictions of the highest degree, which have no significance besides being a good syntactical web in which to catch empirical phenomena.

Let us consider, for example, the model proposed by Rayleigh and Jeans, between 1900 and 1905, to solve the problem of the black body. They considered the black body *as if* it were a cavity whose walls emitted electromagnetic radiation with frequency depending on their temperature. In order to construct the model, they borrowed results belonging to other theories. From classical electromagnetic theory, they took the fact that the radiation inside a cavity had to have the form of stationary waves with nodes at the walls. Geometrically, they derived the number of standing waves within a certain frequency range. Finally, they used the kinetic theory to calculate the average total energy of the waves of the system in thermal equilibrium. By means of these 'pieces' they arrived at an expression for energy density. That is, they constructed a phenomenological model by means of which they hoped to have saved the phenomenon of black body radiation. The end of the story is known: their conclusion did not match with the experimental values. Actually, there was an agreement only for low frequencies, but this was insufficient to consider the model positively. Thus Rayleigh and Jeans's model was abandoned: the pragmatic aims for which it was constructed were not satisfied, the utility of the model having been exhausted, it was simply put aside.

Physical sciences make great use of phenomenological models, as I will show in the discussion of nuclear physics in the next section. Even in biomedical sciences, phenomenological models are widely adopted. Virtually all the mathematical models in population dynamics, epidemiology, population genetics, pathology and physiology, are such.

Let us consider, for example, the well-known Lotka-Volterra model for population dynamics. It happened that, in 1925 the biologist D'Ancona asked Volterra, his father-in-law, whether he was able to construct a mathematical model capable of describing a strange ecological situation: fishing had decreased out of the Adriatic harbours of Trieste, Fiume and Venezia since World War I, but the proportion of predatory fish caught had increased by about 10–12 per cent (in particular those of the class Chondrichthyes, or Selachii). Volterra (1926a, 1926b, see also Volterra 1931) solved this problem by constructing an asif dynamics of population, based on six hypotheses: (1) prey are limited by only one species of predators; (2) in the absence of predators, prey population growths exponentially (in a Malthusian way); (3) the predators eat only one species of prey; (4) there is no competition between predators in finding prey; (5) in the absence of prey, the predators die exponentially; (6) the system prey–predator is close to other biotic and a-biotic influences. Therefore, if p is the size of prey population and P is the size of predator population, then we can write the following set of differential equations:

$$\frac{dp}{dt} = ap - bpP$$

$$\frac{dP}{dt} = -eP + cpP$$

where (1) ap is the prey growth (a is the growth rate, that is, the Malthusian constant); (2) bpP is the prey decrease due to predators (b is the predation rate); (3) eP is the predator decrease due to prey absence or scarcity (e is the predator death rate); (4) cpP is the predator growth due to prey capture (c is the voracity rate). Such a model could be made more realistic by introducing non-linear terms, for example by introducing a term concerning the fishing (let f be the fishing rate). Therefore we have

$$\frac{dp}{dt} = ap - bpP - fp$$

$$\frac{dP}{dt} = -eP + cpP - fP$$

It is clear that the representation offered by Volterra's model, which was also constructed independently by Lotka (1925) to grasp a chemical situation, is fictive. Nevertheless it allowed Volterra to arrive at three laws:

(1) The coexistence of a prey species and of a predator species, under the hypotheses seen, has an infinite oscillatory trend.
(2) The average value of the prey population in a given period is always equal to e/c, and the average value of the predator population in a given period is always equal to a/b.

(3) In the case of fishing, the average value of the prey population grows of the quantity *f/d*, while the average value of the predator population diminishes of the quantity *f/b*.

Moreover, it was able to satisfy D'Ancona's need for a good description and prediction of the empirical situation.[13] In conclusion, it was able to *save the phenomena*.

Let us consider another example: the model describing the variation of a bacteria population. Let $N(t)$ be the number of bacteria at the time t and $N(t+\Delta t)$ the number of bacteria after a time interval Δt. That is, at the time $(t+\Delta t)$ the number is $N(t+\Delta t) = N(t) + \Delta N$. Of course the variation of the population is given by the bacteria which are born minus the bacteria which are dead. If B is the natality constant and D the mortality constant, after Δt we have a population equal to $N = [BN(t)\Delta t - DN(t)\Delta t] = (B-D)N(t)\Delta t$. It means that in the interval Δt there is a variation equal to $\Delta N/\Delta t = (B-D)N(t)$. This equation, for $\Delta t \to 0$, becomes $dN/dt = (B-D)N$. That is, we have $dN/N = (B-D)dt$. The solution of this differential equation is $N = e^C e^{(B-D)t}$, where C is the integration constant. If we call $N_0 = e^C$ the number of bacteria at the initial time t_0, then the variation of bacteria population is described by $N = N_0 e^{(B-D)t}$.

All the above mentioned quantities, that is, N_0, B and D, have a strictly phenomenological significance, in the sense that they can be obtained by empirical and statistical analysis of the population. There is no theory from which the equation of the variation is obtained. It has been constructed, in a very simple way and by using simple mathematics, only *to save the phenomenon* of the variation of the population; nothing more.[14]

An interesting case: nuclear physics. I want to stay for a while with nuclear physics since it wonderfully exemplifies why and how phenomenological models are introduced.[15] For in nuclear physics, unlike what happens in other physical fields, there is no theory (in a strong sense) capable of organizing in a unique and structured framework the enormous amount of experimental data on the atomic nuclei. Instead nuclear physics presents itself as a patchwork of phenomenological models, each one capable of saving phenomena in a particular domain. We know that the atomic nucleus is a very compact system containing most of the mass of the matter in small spaces, and that it is a well-defined system, composed of nucleons (protons and neutrons), whose number goes from A = 1 (Hydrogen) to about A = 238 (Uranium). This number is too low to permit a statistical approach. But at the same time, apart from the lightest nuclei like Deuterium (A = 2), Tritium (A = 3) and Helium (A = 4), it is too high to allow us a detailed and precise solution of the dynamical problem in terms of quantum mechanics.

Even if we do not have a theory able to account for all nuclear phenomena exhaustively and in terms of few interactions among few elementary constituents, we have a set of phenomenological models, each of which can be seen as a satisfactory tool able, within certain well-defined limits, to connect experimental data and to offer good predictions. Every nuclear model, however, more or less coherently refers to a certain set of assumptions constituting what could be called the *nuclear theoretical background* (NTB).

Of course the NTB is not a theory, in the proper philosophical sense, but a collection of statements permitting us both to delimit the investigation field of nuclear physics, and to individualize a reference for all the models. The NTB, enucleated mainly after 1933, can be thought of as made up of two classes of assumptions: (1) those concerning the properties of nuclear objects, (2) those concerning the properties of nuclear interactions. Let us try to schematize them.

Properties of nuclear objects:

(1) The nucleus X is completely identified by the mass number A, with $1 < A < 238$, and by the charge number Z, with $1 < Z < 92$ (the nucleus is composed by A–Z neutrons and Z protons, and therefore its total electric charge is Ze).
(2) Each kind of nucleus, ${}^{A}X_{Z}$, is characterized by a certain number of different properties (the charge, the radius, the mass, the total binding energy, the binding energy of the 'last' nucleon, the reaction and decaying mode, the half-life time, the total angular momentum J, the parity π, the magnetic dipole moment μ, the electric quadrupole moment Q, the level excitation energies, and so on).
(3) The proton mass is $m_p \cong 938.3$ MeV, the neutron mass is $m_n \cong 939.6$ MeV. Protons and neutrons have spin $s = \frac{1}{2}$ and then follow the Fermi-Dirac statistic.
(4) Nucleons are composed of quarks.

Properties of nuclear interactions:

(1) Protons interact by means of electro-magnetic force; nucleons by means of the weak force, the gravitational force and the strong force, which is the dominant one. The Quantum Chromo Dynamics (QCD),

which is known to be the theory of the strong interaction, is unfortunately not yet effective at the level of nuclear application. Therefore, the nucleon–nucleon interaction is tackled by means of phenomenological potentials which implicitly take into account effects due to the exchange of different kinds of mesons.

(2) The nuclear interaction is charge-independent. The neutron and the proton may be considered as different states of the same object (the nucleon) in the sense of the isospin formalism, with reference to the strong interaction effects.

(3) The nuclear interaction has a range of about 1–2 fm (and then it acts at smaller distances than nuclear dimensions).

(4) The nuclear potential is attractive, in the range just mentioned, with an intensity of about 40 MeV.

(5) The nuclear interaction is repulsive for distances smaller than 0.5 fm and it depends on the spin and isospin of the interacting nucleons.

Once the NTB is so defined, we may outline the main approximations involved in the construction of a phenomenological model of a given nuclear situation, even if – of course – different models deal differently with them: (1) the gravitational interaction is completely and always neglected; (2) weak interactions are taken into account only in important but very peculiar cases, like the β decay; (3) electromagnetic interaction among protons cannot be neglected, but it is usually taken into account separately from the nuclear interaction; (4) the effect of the electron cloud on the nucleus is usually neglected and therefore the nucleus is thought of as an isolated system. This is possible because the electron mass is much lower than the nucleon mass ($m_e = 0.511$ MeV, $m_{nucl} = 938$ MeV), and the distance between cloud and nucleus is much higher than the average nucleon-nucleon distance; (5) nucleons are usually thought of as point objects, although the nucleon radius is not fully negligible with respect to the nucleus dimension (the proton radius is $r_p \cong 0.805 \pm 0.011$ fm, and the neutron radius is $r_n \cong 0.36 \pm 0.01$ fm); (6) the nucleus is thought of as a many-body system, treated in non-relativistic approximation, that is, through the non-relativistic Schrödinger equation. Thanks to these approximations, a great theoretical complexity is removed, presumably without losing physical significance. In particular, given any nuclear system, we can formulate its Schrödinger equation:

$$H\psi(\mathbf{r}_1, \mathbf{s}_1, \tau_1, \ldots, \mathbf{r}_A, \mathbf{s}_A, \tau_A) = E\psi(\mathbf{r}_1, \mathbf{s}_1, \tau_1, \ldots, \mathbf{r}_A, \mathbf{s}_A, \tau_A)$$

where

$$H = \sum_{i=1}^{A} \left(-\frac{\hbar^2}{2m} \nabla_i^2 + \frac{1}{2} \sum_{j \neq i} V_{ij} \right)$$

is the Hamiltonian (V_{ij} is the nucleon–nucleon potential and the indexes *i* and *j* refer to nucleons *i* and *j*; *m* is the nucleon mass); ψ is the total wave function of the system ($\mathbf{r}_i$ is the position variable; $\mathbf{s}_i$ is the spin; τ_i is the isotopic spins); *E* is the energy. However, although this equation already represents a notable simplification of the real situation, it is not at all an easy solution. For holding the NTB, and despite the considered approximations, we obtain a Hamiltonian of the nuclear systems which is not resolvable. Here is the reason why it seems impossible to obtain a good nuclear theory. Instead we can construct only a collection of phenomenological models that (1) are more or less coherent with the NTB; (2) start from the considered approximations; and (3) add to these ones other approximations in conformity with the particular formal and empirical necessities to be tackled.

Since the models can be thought of as a sort of bridge connecting the NTB and the experimental data,[16] they can be valued from two different perspectives: (1) from the point of view of the plausibility of the assumptions (that is, with reference to the NTB), and (2) from the point of view of the experimental consequences (that is, with reference to the data). If we neglect the former, we have plenty of freedom of choice of the assumptions (with more consideration given to the mathematical simplification they bring than to their plausibility), and, subsequently, we value the quality and the empirical power of the approximations they introduce. From this perspective, the experimental check constitutes the unique criterion to evaluate the reliability of the assumptions. However, if we neglect the latter, we must be aware that the assumptions could entail only very rough approximations. However, it should be recalled that every approximation is legitimate, if the results of the model coincide with the experimental data within the experimental error.

A last remark. Since all the nuclei are collections of the same kind of particles (neutrons and protons) bound together by the same forces (mainly nuclear interactions), they are completely characterized by the coordinates *A, Z*. By taking this into account, it would be natural to expect an ideal nuclear theory to unify the descriptions of all the nuclei into a unique frame of the type *f(A, Z)* which can be particularized for given values of *A* and *Z*, that is, for a given nucleus.

A typical example of this possibility is given by the case of the binding energy *B*, which is fairly well described by the semiempirical mass formula $B = B(A, Z)$. Unfortunately this formula concerns only one nuclear property and nowadays a more comprehensive *f(A, Z)* does not exist. However, reasoning on *f(A, Z)* may be useful in grasping other peculiarities of nuclear physics. It is not sufficient to consider *A* and *Z* as two parameters characterizing a given nucleus, but rather as parameters characterizing a given domain within which a given model works. For example, (1) the shell model works for *N* and *Z* not very far from the magic numbers; (2) the vibrational model works for $A < 150$ and even *N, Z*; (3) the rotational model works for $150 < A < 190$ and for $A > 230$ with even *N, Z*; (4) the model for γ-unstable nuclei works for $Z > 52$ and even $N < 80$.

At this point I wish to re-state clearly that a model can be accepted, despite its only partial validity, precisely when it is possible to provide empirical and pragmatic reasons for its high organizational and predictive efficiency in a determined domain characterized by a well-defined value of *A* and *Z*, that is, where it is good at saving the nuclear phenomena under inquiry. For example, the shell model deals with single particles and works better next to magic nuclei, where to separate the system into 'core + particle' or into 'core + hole' is possible. This is not possible very far from magic nuclei, as there are too many valence nucleons, and here the model does not work very well. Some purely vibrational models work for nuclei where the electric quadrupole moment *Q* vanishes, and therefore where there is spherical symmetry. Rotational motions are possible only when the nucleus is permanently deformed and then when $Q >> 0$ or $Q << 0$.

However, it should be noted that it is not correct to consider a nuclear model as a rigid, stable and unchangeable structure from which a well-defined set of data can be derived. It is instead more appropriate to regard it as a flexible structure, continually liable to improvements that allow us to account increasingly well for the same experimental data and for new experimental data. Nuclear models are continually modified to increase their efficiency in a given domain. For example, the shell model, thanks to the elaborate formalism of the irreducible spherical tensors, has nowadays reached an organizational capacity and a predictive power much higher than the first formulation. The drop model too has undergone a very high number of changes from its first formulation, substantially immersed in classical physics, to its complex quantum formulation, based on rotation-vibration motions.

In conclusion, I would like to remark again that nuclear physics is a case study particularly suitable for the philosophical comprehension of the status, the structure, the role and the interplay between phenomenological models. It is a matter of fact that a unitary nuclear theory, capable, on the grounds of the NTB, of accounting for the huge amount of experimental data is not available and seems unlikely to be obtainable in the near future. The pragmatic route taken so far by nuclear physicists has been that of constructing a wide plurality of phenomenological models, of high organizational and predictive efficiency in particular domains of phenomena, just to save them.

3.1.5 Models and mathematical schemes

Whenever we speak about mathematization, we must do it carefully. First of all it should be noted that even if we speak of a mathematical model when we rewrite a physical theory with a different mathematics, such a 'model' is completely different from a replacing model of that theory. Under the latter there is the notion of 'material analogy', as clarified by Hesse (1966); under the former there is the notion of 'formal analogy', as stressed by Redhead (1980). Here, the physical theory is 'dressed again' by the new mathematics.[17] Therefore, a 'mathematical model' of a physical theory considered as a 'new mathematical dress' cannot be considered a model in the sense discussed here: it is not a fiction, not an as-if. We do not feign anything when we reformulate a theory by means of new mathematics.

Another source of possible misunderstanding is given by the differential and integral equations of physical and biomedical sciences. It is true that these equations are used to represent many physical situations (from fluidodynamics to celestial mechanics, from electrodynamics to quantum mechanics and so on) and biomedical situations (from population genetics to morphogenesis and so on), but they are not like either a physical fiction or a biomedical fiction, that is, what characterizes a *physical model* or a *biomedical model*. Rather, they are *mathematical schemes*, that is, abstract and void mathematical tools that acquire physical significance or biomedical significance only when they are applied to specific physical or biomedical situations. Without this 'something more', they are nothing but pieces of mathematics. Historically, a mathematical scheme may have been constructed in strict correlation with a model, but its usefulness can overflow into other fields, making it a sort of formal routine for solving structurally analogous problems. For example, the set of differential equations

$$\frac{dx}{dt} = \alpha x - \beta xy$$
$$\frac{dy}{dt} = -\varepsilon y + \chi xy$$

is a mathematical scheme that only becomes significant when α, β, ε, χ and *x*, *y* are interpreted in a certain manner. The scheme can become the Lotka model of a chemical interaction between two reagents, the Volterra model of two interacting populations, but also one of the Turing models of morphogenesis (see Turing, 1952).

The Lagrangian and Hamiltonian formalisms should also be interpreted in this way, that is, as mathematical schemes. Thus, rigorously speaking, it is not correct to talk about the Hamiltonian model of the harmonic oscillator, or the Hamiltonian model of the hydrogen atom, and so on. This way of speaking conceals a misunderstanding because, in fact we are dealing with the application of a mathematical scheme to formalize idealized physical situations. It follows that one should talk about the Hamiltonian of the model of the harmonic oscillator, and the Hamiltonian of the model of the hydrogen atom, and so on. It is the harmonic oscillator and the hydrogen atom that are models, not their Hamiltonian which is the particularization of the general scheme of the Hamiltonian:

$$H(p, q, t) = \sum_i p_i q_i - L.$$

3.1.6 Models and mathematized representations

Another concern is the difference between mathematized theories and mathematized models. This is important, since it allows us to put into a correct philosophical frame many mathematical formulations dealing with biology, economics, meteorology, sociology, the flux of traffic in a great city, the project of a car or a ship and so on.

Let us consider the case of the theory of deterministic chaos (TDC). This allows us to deal with situations that start as completely deterministic, but that subsequently become chaotic and unpredictable. Here the crucial role is played by the initial conditions. If we do not know precisely (that is, without any uncertainty) the initial conditions, we are unable to predict future states of the system. Nevertheless, even if the system is chaotic in the ordinary Euclidean space, the matter is different in the phase space. Here 'strange attractors' appear and allow us to find again, even if qualitatively, the lost predictiveness and the lost order. While in the case of kinetic theory we have a mathematized replacing model of a physical theory, in the case of TDC, we have a mathematized theory allowing mathematized focusing models of situations such as turbulence, crazed compass needle and so on.

In considering this, we should distinguish clearly between two different cases:

(1) we have a mathematized theory, and we can deduce mathematized models from it (as happens in the case of the TDC);
(2) we do not have a theory, and we can construct the model entirely independently of any specific representation, but by borrowing the necessary mathematical or non-mathematical pieces from here and there.

In both cases, we arrive at a mathematized model of a fictive situation, but in the first case we have a focusing model, whereas in the second case we have a phenomenological model.

Let us consider two other instructive examples, that of the thermodynamics of irreversible processes (TIP) and that of the theory of catastrophes (TC). As early as the 1940s, Prigogine began to study what happens in open, non-linear and far from equilibrium chemical–physical systems. He reached a complete representation of these systems. Specifically he found that, even if their internal entropy increases according to the second principle of thermodynamics, the variation of total entropy can vanish, provided that there is a flux of negative entropy from the outside. At a certain point, Prigogine became aware that a huge number of systems might be thought of as open, non-linear and far from equilibrium systems. Therefore, he expanded what had been found for the chemical–physical systems to these new situations. In such a way he thought he had found 'a new alliance' (see Prigogine and Stengers, 1979) between the world of nature and the world of life. By means of this approach, he was able to represent almost everything: from Bénard vortexes to the Zhabotinski reaction, from the chemical reactions among molecules to the organization of a colony of termites, from the growth and collapse of an ecosystem to pre-biological evolution, to social-cultural development, and so on. Wherever there was the possibility of representing something as an open system, there Prigogine's approach struck. Ignoring the hubris of such holism, attention should be paid to the fact that the models for these cases are focusing models because they are derived from a theory, the TIP.[18]

Unlike the TIP, the TC, proposed by Thom, arose as a mathematical theory describing the drastic passage of a system from an evolution governed by one set of differential equations to an evolution governed by another set of differential equations. TC has also been discussed at length, and many people have seen it as a sort of formal panacea able to describe an incredible quantity of phenomena – physical, biological, social, historical and so on. Thus TC has become a representation for every kind of transition. In this case there is a mathematized abstract and powerful scheme allowing us to construct mathematized phenomenological models able to fictionally represent as-if worlds, where there are physical bifurcations, revolts in the jails, biological mutations, historical or cultural revolutions and so on. Summing up, the models realized inside the TIP are focusing models, while the model realized through the mathematical schemes offered by TC are phenomenological models.

3.1.7 Ideal-typical models

We have considered classes of models whose aim is to organize already known data and to predict new data by starting from false premises. They are fictive representations devised for practical aims. But there is another class of fictive representations devised for practical aims, even if they neither organize already known data or predict new ones. These are those models, and therefore fictive representations, which are used as ideal-typical as-if situations. It was Weber, in his great paper of 1904, 'Objectivity in Social Science and Social Policy', who outlined the main characteristics of this class of models. They are fictive representations of real situations (peasant economy, artisan economy, capitalistic economy and so on) constructed to understand the real situations by comparing them with the fictive representation through the emphasis on the differences between the two.

Extremely close to the Weberian class of ideal-typical models, are those discussed by Popper (1945, 1957, 1972) in his 'logic of situation'. Here Popper suggests constructing idealized models, that is, ideal-typical representations, of real situations to understand the rationality of real agents. In this case, Popper proposes realizing an ideal-typical model of rationality to judge the value of real actions performed by real agents.

This is more or less what Durkheim (1895) had already suggested. He even wrote that we should create a new branch of sociology concerning the construction of these ideal-typical models. With reference to ideal-typical representations, we should, however, be careful to distinguish between ideal-typical models as sets of behavioural norms to be followed to obtain a certain goal (as happens in case of the prescriptivistic methodology) and ideal-typical models as fictive representations of situations with which real situations must be compared to be understood. The Hardy-Weinberg model in population genetics offers a very good example of this kind of ideal-typical representation. Let us consider a population, and the case in which only two alleles A and B sit at the same locus. In this case, there can be individuals that are

(1) homozygotes AA (and let x be the frequency in the population of the genotype AA),
(2) homozygotes BB (and let z be the frequency in the population of the genotype BB),
(3) heterozygotes AB (and let y be the frequency in the population of the genotype AB).

Therefore if p is the frequency in the population of the allele A, and q is the frequency in the population of the allele B (with $p+q=1, p=x+\frac{1}{2}y$, $q=z+\frac{1}{2}y$), then we arrive at the statement that the allele frequencies p and q are constant from generation to generation, and from the first filial generation on we have $x=p^2$, $y=2pq$, $z=q^2$.

Nevertheless, to obtain this result we must construct a fictive, or an as-if population, that is, a model of population, by imposing the following hypotheses: (1) there are only diploid organisms; (2) there is sexual reproduction; (3) the sex ratio is genotype independent; (4) fertility is genotype independent; (5) survivorship is genotype independent; (6) there are non-overlapping generations; (7) there is random mating; (8) the population is very large; (9) there are equal allele frequencies in the two sexes; (10) there is no migration; (11) there is no mutation; (12) there is no selection.

As is clear, these hypotheses depict a totally fictive world, in particular an ideal-typical world, which is used by biologists to understand real situations by comparing the genotype frequencies in real populations with the Hardy-Weinberg frequencies. For, from possible deviations, considerations about, for example, selection, or non-random mating can be inferred.

3.2 Thought experiments

So far, I have considered theories and models, that is, respectively, hypothetical and fictive representations having empirically testable consequences. However within the scientific domain there are also thought experiments and they can be framed in the same way as the fictive approach discussed above. To tackle this question, I will begin with the analysis of a well-known historical case.

3.2.1 Newton and Mach: a case of rhetorical thought experiment

At the end of the Introduction to his *Principia*, Newton (1687) discusses the possibility of showing the effects of absolute motion and thus the real existence of absolute space. In order to achieve this, he proposes the experiment of the rotating bucket and the experiment of the two rotating spheres. The first consists in taking a bucket full of water and hanging it on a wire. Then the bucket is rotated until the torsion of the wire is completed. When it is freed, the bucket rotates in the opposite direction, and the water inside rises at the side and lowers at the centre. For Newton, the rising, or the lowering, confirms the presence of absolute motion, and thus of absolute space. In this case we have more than a fiction. The fiction represents an as-if world composed of nothing but the bucket and the water. It is a typical case of a fiction derived from a theory, that is, classical mechanics. Within this fiction, centrifugal force can be found. Yet Newton does not limit himself to proposing this fictive world, rather he assumes it as the starting point for an argument aimed at showing the existence of absolute space. In this way Newton is proposing precisely what I call a *rhetorical thought experiment*. From this first example, the main feature of this kind of thought experiments can be grasped: *there is an argument starting from an as-if world, and arriving at supporting a non-empirically checkable thesis.*[19]

Of course as there are arguments for and against a thesis, so there are rhetorical thought experiments for and against a thesis. A paradigmatic example of this second kind of rhetorical thought experiment is given by the argument that Mach (1883) addressed to the Newtonian conclusion just seen. Mach[20] affirms that he would agree with Newton's conclusion if the bucket stayed motionless and the rest of the universe rotated. In this case, if the surface of the water stayed flat, we could conclude that absolute motion exists. However, this cannot happen, or there would be two different behaviours within the same universe: the first one due to the rotation of the massive celestial bodies, the second one due to the rotation of the bucket.[21] Here, as in the Newtonian case, we have a rhetorical thought experiment: there is a fiction, representing an as-if world full of matter, from which an argument starts. But now it is addressed against a thesis, in particular Newton's.

Fallacious rhetorical thought experiments

The above analysis has led to the interpretation of an important class of thought experiments as arguments starting from fictive representations of an as-if world. However, an argument may be fallacious and consequentially a rhetorical thought experiment may also be fallacious. A good thought experiment is a good argument, whereas a fallacious thought experiment is a fallacious argument.[22] Now, an argument is fallacious when either the premises are weak, false or incomplete, or the inferences used to reach the conclusions are wrong.

Let us return for a while to the Newtonian thought experiment. Newton's fiction is completely correct. The removing of the water from the centre is a perfectly legitimate centrifugal effect that shows the presence of a privileged frame. Yet affirming the physical–mathematical existence of a privileged frame is quite different from affirming that a real absolute frame exists. It is enough to state that, in that particular representation of fictive aspects of the world, there is a frame with certain formal properties. The worm in the Newtonian argumentative reasoning lies in the fact that he moves from a fictive representation of the world, where there is a fictive privileged frame, to a representation of the world infected by naive realism, where the fictive privileged frame becomes an absolute frame, in a metaphysical sense. Therefore, there are no errors in Newton's physics, that is, the premises of his argument for absolute space are valid, but there is a fallacy in the argumentative inferences from the starting fiction to the conclusions.[23]

Whereas the Newtonian thought experiment is an example of a fallacious thought experiment due to the fallacy of the argumentative inferences, the Einsteinian thought experiment against Heisenberg uncertainty relations is a good example of a fallacious thought experiment due to wrong premises. At the 1930 Solvay Meeting, Einstein proposed an argument starting from a representation of a fictive world composed of a box full of radiation with a hole which was closed and opened by a clockwork mechanism. In this way, the precise time of the photon's leaving the box can be measured. Moreover, if the box is weighed before and after the emission of the photon, its precise energy can be determined, by means of the well-known equation $E = mc^2$. Therefore, both the precise time and the precise energy can be calculated, and in this way Heisenberg relations are infringed. Unfortunately this thought experiment is wrong, as Bohr (1949) argued, because Einstein forgot to take into consideration a necessary premise that, ironically enough, concerned his general relativity.[24] That is, he did not propose a fallacious argument, but he assumed a physically wrong as-if world.

3.2.2 Exploratory/clarifying thought experiments

Besides the rhetorical thought experiments, there are also exploratory/clarifying thought experiments. These are thought experiments used to explore the regions of a theoretical domain that are not completely clear, or to clarify what a certain concept, or a certain set of concepts, means, and to what consequences it can lead. They are conceptual tools by means of which we learn something of our concepts and of our world (see Kuhn, 1964). This class of thought experiment is different from the one just discussed, because it does not contain any trace of rhetorical argument for or against a thesis, but only a reflection on what a particular as-if world involves.[25]

Let us consider two thought experiments. The first is the well-known thought experiment suggested by Einstein et al. (1935) in terms of wave equation, and reformulated by Bohm (1951, pp. 614–19) in terms of spin functions, to argue against the completeness of quantum mechanics. The second thought experiment is discussed by Einstein (1911) in the context of classical gravitational theory, to show that there is a red shift in the frequency of a photon moving in a gravitational field. Einstein proceeds as follows.[26] A free falling particle of mass m is going from point A to a lower point B distant h. At B, the mass of the particle is $M = m + mgh$ owing to the kinetic energy it acquires. We suppose that at B the massive particle is transformed into a photon with energy M moving upwards to A. If, during the ascent, the photon did not interact with the gravitational field, at A it would always have the same energy M. If, at A, it were retransformed into a massive particle this latter would have mass M. That is, we would have an energy surplus equal to mgh produced from nothing. But this is not possible. In order not to infringe the principle of the conservation of energy, the photon, during its upward motion, has to lose energy. Thus its frequency has to diminish step by step. That is, if the photon is an optical one, its frequency shifts towards the red.

In the first case we have a rhetorical thought experiment, by means of which Einstein et al. wanted to argue against orthodox quantum mechanics by showing the paradoxical consequences to which it led.

In the second case, we have a thought experiment by means of which Einstein explores a theoretical physical possibility. In both examples, there is a fictive world within which fictive mechanisms work. Nevertheless, *in the first one the fiction is the basis from which an argument starts, whereas in the second situation there is a reflection on what an as-if world involves.*

At this point a question could be posed: what is the difference between models and exploratory/clarifying thought experiments, if both are fictions?[27] The model, as I stressed, even if it is a representation of an as-if world, can be empirically tested by means of its consequences. A model is empirically powerful. It does not matter if a model is constructed by diverging from the facts, because what is relevant is the particular problem it must solve and, at the end, that its consequences correspond to the facts. This is no longer true for the exploratory/clarifying thought experiments which do not admit empirically testable consequences. This kind of thought experiment is not proposed to satisfy a pragmatic need, and it does not lead to any empirically testable consequences. Instead it is constructed to explore theoretical possibilities and/or to clarify their power.

Metaphorically, we could say that whereas models have their heads in the clouds but their feet on the ground, exploratory/clarifying thought experiments have both head and feet in the clouds, as is wonderfully exemplified by the model that Malthus proposed in his 1798 *Essay on the Principle of Population*, and that strongly influenced Darwin. Malthus's idea, according to which a population, when unchecked, increases in geometric ratio, while subsistence increases in arithmetical ratio, is based on the following assumptions: (1) the population must be so large that stochastic effect can be neglected; (2) the population is censused each t years;[28] (3) there are no competing species; (4) there are no intra-species competitions; (5) the death rate and the birth rate are constant; (6) the death and the birth rate do not depend on the size of the population; (7) there are no a-biotic influences (no geophysical environment dependences, no wars, no food dependences and so on). In this case, given r the reproduction rate per year, d the mortality rate per year, and N_t the number of individuals in the t-year, we can write

$$N_{t+1} = N_t + rN_t - dN_t = (1 + r - d)N_t = aN_t$$

that is,

$$N_{t+1} = aN_t$$

which is called the Malthusian equation; a is the constant Malthusian growth rate. Therefore, if $N_{t=0} = N_0$, the solution is

$$N_t = N_0 a^t.$$

As should be clear, by means of Malthus's thought experiments, we can explore a theoretical situation, in particular what would occur if a species could increase in number without any kind of limits, but we have no real possibility of checking it empirically.

Even if thought experiments can be divided into two different classes, I am not claiming that a rhetorical thought experiment cannot become an exploratory/clarifying thought experiment, and vice versa. Both of them are empirically impotent and both of them contain a fictive representation of the world. The as-if world depicted in an exploratory/clarifying thought experiment can become the basis for an argument for or against a thesis, and therefore it can turn into a rhetorical thought experiment. Vice versa, a rhetorical thought experiment deprived of its rhetorical aspects can become an exploratory/clarifying thought experiment. This latter case is emblematically exemplified by the Einstein et al. thought experiment which, at the beginning, was proposed as a rhetorical thought experiment but then was transformed, especially by Bohm (1951), into an exploratory/clarifying thought experiment. Now, only the fictive construction was considered and the argumentative part was neglected.

3.2.3 Thought experiments and real experiments

According to some authors, in particular Mach (1905), a thought experiment is a sort of project preceding real experiment. That is, first the thought experiment is proposed and then the real experiment is carried on. In fact, the relation between the two is not so simple. Let us try to clarify this point by means of some examples.

Most of the thought experiments proposed in the debate on the foundations of quantum mechanics can be considered as idealized projects which can stimulate real experiments. This is the case, for example, of the Einstein-Podolsky-Rosen-Bohm thought experiment, which (of course deprived of the rhetorical aspects) can be considered as an idealized project which has stimulated the realization of a real experiment. Incidentally, it is no coincidence that the authors of one of its realizations have emphasized this connection even in the title of their paper: 'Experimental realization of Einstein-Podolsky-Rosen-Bohm *Gedankenexperiment*: a new violation of Bell's inequalities' (see Aspect et al., 1982).

Nevertheless there are thought experiments in the foundations of quantum mechanics, for example those exploratory/clarifying thought experiments contemplating wormholes (see Gonella, 1994), which can never stimulate any real experiment for the simple reason that to arrange a suitable experimental apparatus is impossible.

Both classes deal with fictive experiments realized inside a fictive world, but they differ in some aspects. Whereas the first fictive experiments can be more or less easily transformed into real experiments by arranging a suitable experimental apparatus, the second ones cannot, since to set up the apparatus is technologically, or theoretically, impossible. Therefore we are forced to conclude that not all thought experiments can be considered as the starting point for real experiments. Moreover, even if some thought experiments could be realized, sometimes such experimentation can be theoretically and experimentally useless. For, what is important in many thought experiments is not to obtain real experimental results, but to understand the theoretical consequences of the as-if world in question. It follows that there are realizable thought experiments which are not worth realizing, owing to the insignificance of the empirical results to which they would lead.

With reference to the relations between thought experiments and real experiments, we should not confuse real experiments with the realization of what I call *as-if experiments*, that is, with the actualization of certain exploratory/clarifying thought experiments. In order to understand this, consider Miller's famous experiment in 1953.

It is well known that the contemporary debate on the origin of life may be dated back to the seminal works by Oparin (1924) and Haldane (1929). Oparin proposed that in suitable situations a solution of proteins can lead to vesicles endowed with metabolism. Haldane, by focusing on viruses, suggested life began with something similar to them, that is, with something similar to RNA. Summing up, Oparin thought that proteins and metabolism were first; Haldane thought that RNAs and replication were first. Who was right? At that time it was simply impossible to settle the quarrel, not least because both Oparin's and Haldane's ideas were theoretical speculations based on suppositions on what the primordial Earth was like. The matter changed in 1953, when Miller proposed his as-if experiment, to settle the question.

Miller, following Urey's hypothesis on the primordial atmosphere, realized one among the infinite possible mixtures of ammonia (NH_3), methane (CH_4), hydrogen (H_2), and water (H_2O). This mixture was thought of *as if* it was the primordial atmosphere. Then, he connected it with boiling water, thought of *as if* it was the primordial ocean heated by the Sun's rays. By means of electrodes he energized the gas mixture, and the electric discharges were thought of *as if* they were lightning. After a couple of weeks, he obtained hydrogen cyanide and formaldehyde. These underwent further reactions in aqueous solution and, in the end, some amino acids (above all glycine, alanine and aspartic acid) were obtained.

The result was surprising and exciting but there was (and there is) a problem, as Miller himself reported: 'Nobody questioned the chemistry of the original experiment, although many have questioned what the conditions were on the prebiotic Earth. The chemistry was very solid' (Henahan, 1996, p. 5). Therefore the problem is: 'What were the conditions on the pre-biotic Earth?' Let us consider this aspect.

Let us suppose we know the exact conditions between 4×10^9 and 3.5×10^9 years ago, and let us say that they are given by the set $< A^1, \ldots, A^n >$. How many hypotheses on the origin of life are compatible with this set? They are infinite: we are in a case of underdetermination of hypotheses by data. Unfortunately, to complicate the matter still further, we do not even know that set. In fact we have an infinite number of possible sets: $< A^1, \ldots, A^n >_1, \ldots, < A^1, \ldots, A^n >_\infty$. But if any set allows us infinite hypotheses on the origin of life, we are in a troubling position. To be honest, we do have a window of possibilities: we know, more or less, the range of pressure, of temperature, of pH and so on. However we still have an incredibly high number of empirically plausible hypotheses. If this holds, a new problem arises: 'How can we perform an experiment based on an empirical situation that we know very roughly?' Actually we can realize an *as-if experiment*. We construct an empirical situation *as if* it were the real one, even if we know that it was not necessarily so. Then we try to understand what happens in that as-if situation. In such a way, we physically realize an exploratory/clarifying thought experiment, that is, an as-if experiment. And this must not be confused with a real experiment.

3.2.4 What thought experiments are

Through the example above, I have illustrated what thought experiments are and what they are not. Now I will summarize the conclusions.[29]

(1) All thought experiments are empirically impotent constructions, that is, without any empirically testable consequences, based on representations of as-if worlds.

(2) There are two different classes of thought experiments: exploratory/clarifying thought experiments and rhetorical thought experiments. The thought experiments of the first class are constructed for theoretical aims, being conceptual tools by means of which certain theoretical questions are explored or clarified by reflecting on as-if worlds. The thought experiments of the second class are used for rhetorical aims. They are characterized by an argument starting from an as-if world and by means of which a thesis is supported or objected.

(3) There are exploratory/clarifying thought experiments that (a) can never be transformed into real experiments; (b) can be considered idealized projects of real experiments, although it is theoretically and empirically useless to realize them; (c) can be considered idealized projects of real experiments, whose realization has great theoretical relevance; (d) can be realized as as-if experiments.

3.2.5 Conclusions on fictive representations

If what is proposed works, models and thought experiments may be framed in the same philosophical conception, based on Vaihinger's fictionalism. For *models* (made to satisfy pragmatic aims) are fictions with empirically testable consequences; *exploratory/clarifying thought experiments* (made to satisfy theoretical aims) are fictions without empirically testable consequences; *rhetorical thought experiments* (made to satisfy rhetorical aims) are arguments starting from fictions.

3.3 Fictions and counterfactuals

In this chapter, I have discussed both models and thought experiments on the basis of the notion of 'fiction'. Needless to say, a fiction is nothing but a counterfactual, and we know that over these last years, a number of philosophers have tackled this topic. Nevertheless, it is interesting to note that some of them seem to discuss an issue never considered before. In fact the problem of counterfactuals has been extensively discussed in the past, even if not under this label, as, for example, happened in the Roman age with the concept of *fictio juris*. However, in what follows I do not want to offer a historical review of the problem, but to look in the history for fresh ideas capable of solving it without resorting to the logico-linguistic techniques that appear to have failed. In so doing I will develop a *hermeneutical approach*, which will enable me to propose a new perspective on counterfactuals.

3.3.1 A well-known analytical$_{l\text{-}l}$ history

The classical problem of counterfactuals, that is, the problem of understanding what the conditions of truth of a counterfactual proposition are, was – and it could not be otherwise – an insuperable obstacle for the neopositivistic philosophers. There are two reasons for this. The first concerns the empirical world: as the antecedent of a counterfactual goes against the facts, there cannot exist any simple or complex fact to which the counterfactual proposition can be reduced. The second regards that the truth of a counterfactual proposition cannot be a function of the truth of its components, owing to the classical definition of the material implication.

Many solutions were proposed to this *impasse*: metalinguistic solutions, pragmatic solutions, modal solutions and so on. Whatever the value of the solutions given by, for example, Goodman, Stalnaker, Lewis and so on, it is historically and theoretically worth inquiring if anyone not belonging to what could be called the *logico-linguistic analytical tradition* (*analytical$_{l-l}$ tradition*) has arrived at a satisfactory solution to the topic.

Taking into consideration that philosophy is not just analytical$_{l-l}$ philosophy, it is plausible to suppose that there are non-analytical$_{l-l}$ thinkers who tackled the topic, very probably using different methods, different terminology and identifying the problems in different terms. This supposition reveals itself to be true as soon as we read, for example, works by Weber and Vaihinger. It is precisely by studying these authors that we become aware that counterfactuals can be faced not only from the point of view of an analytical$_{l-l}$ approach, but in another different and promising way. This is characterized by a strong emphasis on the hermeneutical conditions for understanding the value of the counterfactual statement, but for all this it is no less analytical than the logico-linguistic approach. This is why I call it an *analytical hermeneutical approach* (*analytical$_h$ approach*).

3.3.2 A less-known analytical$_h$ history

Weber's judgement of possibility

In 'Critical Studies in the Logic of the Cultural Sciences', a paper written in 1906, Weber criticized the thesis presented by Meyer in his 1902 *Zur Theorie und Methodik der Geschichte*. According to Meyer, to enquire what would have happened if the fact Y had occurred instead of the fact X leads to insoluble and idle questions. Opposing this thesis, Weber, after stating that not all insoluble questions are idle (as the problem of the existence of God teaches us), shows how 'the judgement of possibility', in other words a *historical explanatory/clarifying thought experiment*, or a *historical counterfactual*, plays an all-important and necessary role in historical research.

Suppose that at the time T the event E took place. The question a historian must solve concerns what caused it. In order to reply, according to Weber, the historian must formulate judgements of possibility, namely, propositions on possible historical (causal) scenarios. In other words, the historian must let his mind move from the actual facts which happened at $t < T$ to possible facts, to build up a 'fantastic picture' of what happened from t to T, that is, to use Vaihinger's jargon, to construct a fictive historical development, an as-if historical development.

Of course, what happened at $t < T$ is infinite, and not all of them can be responsible for the causation of E, but even the facts that could have caused E are infinite. To reduce such infinities the historian, for Weber, must adopt exactly the method considered idle by Meyer.

Specifically, we begin by hypothesizing that at $t < T$ an actual fact, or a set of actual facts, is changed or is absent. Starting from such a 'fantastic' situation, we analyse the historical development from t to T. If the historical development led to an event $E' \neq E$, then that changed or absent actual fact (or actual set of facts) would have a historical relevance, since its non-happening, or its happening in a changed way, would have led to a state different from the state which actually took place at T. On the contrary, if, in the 'fantastic' reconstruction of the historical development, E happened notwithstanding the altered situation with modified initial facts, or set of facts, that initial fact, or set of facts, would be without any historical relevance.

This means that if at t the set of historical facts was $(A_1, \ldots, A_n)$, to understand if the A_i is relevant to the causation of E at T, the historian must answer the following question:

'If, at t, A'_i had happened instead of A_i, what would have happened at T?'

There can be two answers:

(1) 'If, at t, A'_i had happened instead of A_i, then E′ would have happened instead of E.'
(2) 'If, at t, A'_i had happened instead of A_i, then E would always have happened.'

In the first case A_i is causally relevant to the happening of E, in the second case it is not. At this point, according to Weber, the problem no longer concerns the grasping of the historical value of a given event A_i, but the grasping of the epistemological value of the judgements of possibility in cases 1 and 2. Therefore, it is from this epistemological judgement of the proposition that the correct historical judgement on the fact A_i is made. That is, the new problem concerns the validity of the historical counterfactual that has permitted the historian to attribute a certain causal role to an event at $t < T$.

Before tackling this question, however, Weber points out that he is not the first author to propose what he calls a theory of 'objective possibility', namely, of counterfactual argument. For, he notes, a similar doctrine had already been put forward in 1888 by the physiologist von Kries in a paper entitled 'Über den Begriff der objectiven Möglichkeit und einige Anwendungen desselben', and by criminologists and jurists such as Liepmann, Radbruch and von Rümelin. With reference to this historical note, let us observe that it is obvious that the jurists' job deals with such things, as the Roman jurists knew: every day they have to solve questions regarding causal attribution, that is, 'if and when the objective, purely causal, attribution of the consequence of an agent's action is sufficient to qualify such an action as his own subjective fault' (Weber, 1906). Therefore, as mentioned above, the new problem is to grasp the validity of a judgement of possibility such as 'If A had happened, then E would have happened.' That is, the problem is to grasp how the historian, or the jurist, can judge the validity of a counterfactual. To solve this problem, Weber points out that whoever values the judgement of possibility must possess two kinds of knowledge: (1) *ontological knowledge,* that is, knowledge concerning the historical facts, which is obtained by studying historical sources and documents; (2) *nomological knowledge,* that is, knowledge related to the basic rules according to which a human agent 'usually' behaves.

So, the social scientist 'imaginarily' modifies the actual facts occurred at *t*, and '*proceeds to an unreal construction*' of a possible scenario (that is, an as-if scenario). In other words, he depicts a 'fantastic picture' of what happened at $t < T$ and develops it according to the 'usual' rules of human behaviour to see what happens at *T*.

It should be noted, as Weber emphasizes, that the final judgement, even if it is a judgement of possibility, must not be understood as a judgement which reveals our subjective ignorance, but as a judgement clarifying a possible historical development on the basis of our ontological and nomological knowledge. From this point of view, the historical method based on the judgement of possibility, contrary to Meyer's thesis, must be considered to be valid to grasp the relevance of a causal attribution.

Vaihinger's fictive judgement

Vaihinger's approach to counterfactuals is a result of the fact that he wants to explain both linguistically and epistemologically the meaning of 'fiction', that is, the notion at the basis of his theoretical proposal. In chapter 12 of *The Philosophy of 'As If'*, he divides the propositions into 'primary judgements', that is, the assertive or categorical propositions that he considers 'saturated propositions'[30] because they state that 'A is B', and 'secondary judgements', obtained from the primary ones by altering the saturation. Among the latter we have the 'secondary negative judgements', in which the alteration of the saturation is obtained by denying the second terms ('A is not B'), and the 'secondary problematic judgements', in which the saturation is 'variable' because we state that 'Perhaps A is B' or 'Perhaps A is not B' (§§ 26–27).

Among the secondary judgements, Vaihinger also inserts the 'fictive judgements', that is, propositions such as:

'A is to be considered *as if* it were B'.

For these propositions involve an alteration of the saturation of the primary judgements. However, as Vaihinger remarks, the fictive judgement is extremely atypical because at the same time it expresses, first, *the negation of an objective validity*; in other words, it states that 'A is not B'. Therefore the fictive judgement decrees the unreality, or the impossibility that A is B. Second, it expresses *the assertion of a subjective validity*; that is, that A is B only fictionally and for practical aims such as those allowing the knowing subject to find his bearings and live in the world around him by acting in a certain way. Summing up, the fictive judgement does not express any theoretical truth but it does express a pragmatic truth related to the intention to act in the real world. Accordingly propositions such as:

(a) 'If man were free, then he should be judged accordingly.'
(b) 'If the circle were a polygon, it should be submitted to the laws of straight figures.'
(c) 'If a diamond were a metal, it would be fusible.'
(d) 'If Caesar had not been killed, he would have been Emperor.'

express 'absolutely necessary relationships between what is unreal or impossible'. For, even if in the first part of the fictive judgement an unreality, or impossibility, is expressed, from such an unreality, or impossibility, the second part of the fictive judgement necessarily follows.

It is not by chance that Vaihinger exemplifies his analysis by means of a typical Kantian proposition: 'The behaviour of man has to be judged as if man were free.' According to Vaihinger, from an objective point of view, we know that man is not free, but with reference to the pragmatic aims of judging human actions, that is, from a subjective point of view, we pretend that man is free.

The result is that the fictive judgement cannot be a simple case of 'A is not B' because this statement does not reveal any connection between the antecedent and the consequent. However, a relation does exist, and we must express it. In other words, according to Vaihinger, the fictive judgement (the counterfactual proposition) is the ultimate synthesis of an entire argument to be clearly and fully expressed. That is, the fictive judgement

> 'The behaviour of man has to be judged as if man were free'

is the synthesis of the following fictive argument:

Theoretical premises
(1) 'I know that man is not free.'
(2) 'It necessarily follows that we do not have to judge him as if he were free.'

Practical premises and conclusion
(3) 'However, I feign that man is free.'
(4) 'In such a case, it necessarily follows that we can judge him accordingly.'
(5) 'Therefore, the laws according to which actual man behaves are fictionally (and for practical aims) assimilated to the laws which necessarily follow from the unreal or impossible existence of free men.'

Thus, a correct analysis of the fictive judgement 'If A were B, then C' (for example, 'If Caesar had not been killed, he would have been Emperor') can be made if, first of all, the fictive argument of which it is a synthethis is expressed in a clear and full way. Then, the fictive argument must be split into a theoretical part, which is objectively valid, and a pragmatic part, which is subjectively valid. Specifically,

Theoretical part (objectively valid)

1. 'A is B' is false.	1. 'Caesar was not killed' is false.
2. Then, consequently C is false.	2. Then, consequently, 'Caesar had been Emperor' is false.

Pragmatic part (subjectively valid)

3. Anyhow, I feign that 'A is B' is true.	3. Anyhow, I feign that 'Caesar was not killed' is true.
4. Then, consequently C is true.	4. Then, consequently, 'Caesar had been Emperor' is true.
5. In this case, C is/was/will be a property/quality/future state of A.	5. In this case, being Emperor was a future state of Caesar.

What Vaihinger tries to do is to recover at a pragmatic level what theoretically is simply false. So, firstly a fictive and imaginary (we would say 'counterfactual') world, that is, an as-if world, is constructed so that it develops according to the same rules of the actual world. Then, after constructing the as-if world and after deducing consequences by means of the necessary and actual rules, such consequences are moved into the actual world so that the agent may find his bearings and live.

However, one point should be taken into consideration. According to Vaihinger, the main difference between the actual world and the fictive world does not lie in the rules regulating the development of the two worlds, but in the fact that the fictive world has no theoretical value, only a pragmatic value.

3.3.3 Counterfactual proposition and counterfactual argument

What Weber and Vaihinger are both tackling, even if with different languages and for different aims, is not the problem of the truth of a single counterfactual proposition but the problem of the validity of the entire counterfactual argument expressing in a clear and full way the content of the connected counterfactual proposition. Both authors suggest that underneath a judgement of possibility (Weber), or underneath a fictive judgement (Vaihinger), that is, at the core of a counterfactual proposition, there is a complete argument that cannot be overlooked. In what follows I will complete and improve the legacy left by these two analytical$_h$ philosophers, and so suggest an approach, different from the usual analytical$_{l-l}$ one, which should allow us to tackle positively the question of how to grasp counterfactuals.

Accepting part of Vaihinger's analysis, the starting point of my proposal is to express in a clear and full way the counterfactual argument synthesized by the counterfactual proposition. That is, a given *counterfactual proposition*

'If it were (or had been) A, then it would be (or would have been) C'

has to be thought of as the *synthesis of a counterfactual argument* of the following type:

(1) a is A;
(2) There exists a hypothetical rule R which, as far as we know, is valid for past, present and future cases where there is an x belonging to A;
(3) A proposition on a – that is, P_a – follows from R;
(4) It is supposed that a' is an x belonging to A, then a' is hypothetically a case contemplated by R;
(5) Therefore, the proposition on a' – that is, $P_{a'}$ – follows from R.

Let us consider the following counterfactual proposition:

'If we let (or had let) a stone drop, it would fall (or would have fallen) to the ground.'

This has to be thought of as the synthesis of the following counterfactual argument:

(1) The stone (a) is a heavy body (A);
(2) There exists a hypothetical rule R, in this case the law of gravity, which, as far as we know, is valid for past, present and future cases where there is an x which is a heavy body (x belongs to A);
(3) The fall towards the Earth of a heavy body (that is, a proposition on $a - P_a$) follows from the law of gravity (R);
(4) It is supposed that an object (a') is an x belonging to A, then a' is hypothetically a case falling under the law of gravity (R);
(5) Then, the fall of the heavy object a' (that is, the proposition on $a' - P_{a'}$) follows from the law of gravity (R).

The same philosophical analysis has to be made with historical counterfactual propositions such as:

'If Charles Martel had not defeated the Arabs at Poitiers, they would have conquered France.'

For,

(1) At Poitiers the Arabs (*a*) were defeated (non-*A*; where *A* = to be not defeated);
(2) There exists a hypothetical rule *R*, in this case the trivial one according to which those who are not defeated continue to go on, which, as far as we know, is valid for past, present and future cases where there is an *x* which is not an army defeated during a battle (*x* belongs to *A*);
(3) The fact that the defeated army at Poitiers does not go on (that is, a proposition on $a - P_a$) follows from the rule according to which those who are not defeated continue to go on (*R*);
(4) It is supposed that a historical fact (*a′*) is an *x* which is a non-defeated army (*A*), then *a′* is hypothetically a case contemplated by the rule according to which those who are not defeated continue to go on (*R*);
(5) Then, the going on of the Arab army (that is, the proposition on $a' - P'_a$) follows from the rule according to which those who are not defeated continue to go on (*R*).

3.3.4 The judge's hermeneutical region

The weak point of Vaihinger's proposal, as described above, is that he does not accurately analyse a very important aspect of the counterfactual question, that is, what 'the consequent *necessarily* follows from the antecedent' means. This weakness is not present in Weber's approach, as the discussion of nomological knowledge reveals.

Such nomological knowledge is extremely important because only by means of this information are we able to judge if, and why, the consequent really follows from the antecedent. However, it should be noted, as many authors have pointed out, that nomological knowledge is not always made up of rules with the same epistemological status. For, when we tackle scientific counterfactuals such as, 'If the atom were a solar system, then the electrons would be the planets', or 'If we let a stone drop, then it would fall to the ground', the nomological knowledge is made up of rules with an epistemological status quite different from the epistemological status of the rules regarding nomological knowledge by which we tackle, for example, a historical counterfactual such as, 'If Napoleon had not embarked on the Russian campaign, then he would have been the unopposed ruler of Europe.' However, this problem is not at the focus of my interest here. Besides nomological knowledge the agent judging the validity of the counterfactuals must also possess what

Weber calls ontological knowledge. That is, knowledge related to the facts concerning the domain of that particular counterfactual.

Moreover, nomological and ontological knowledge are not sufficient to judge the validity of a counterfactual. If, as shown, it is not true that we have to tackle a single counterfactual proposition but an entire counterfactual argument, we must possess knowledge concerning the way in which we infer from one proposition to another within such an argument.

Therefore, besides *nomological knowledge* and *ontological knowledge*, a judge must possess also *inferential knowledge*. That is, knowledge as to how argument develops in that domain.

Briefly, the judge achieves his judgement of validity by applying to the counterfactual what I call a *hermeneutical region* belonging to his background knowledge. And this *hermeneutical region* consists of (1) the nomological part, (2) the ontological part, and (3) the inferential part. At this point a question arises: 'How does the judge choose the right portion of his background knowledge to use?' That is, how does he choose the hermeneutical region?

The adequacy of the hermeneutical region and the analytical$_h$ problem of counterfactuals

It should be clear that *we are not dealing with counterfactual propositions but with counterfactual arguments*, of which counterfactual propositions are the synthesis. We are before an iceberg whose visible tip is given by the single proposition whose *raison d'être* is given by the immersed and non-visible part, namely, the whole counterfactual argument. It is exactly such counterfactual argument as a whole that must be judged. This is made possible by the judge's application of a hermeneutical region belonging to his background knowledge. Therefore, the judgement results from the *hermeneutical comparison* between the applied hermeneutical region and the supposed counterfactual argument synthesized by the counterfactual proposition.

On the one hand, there is the counterfactual proposition (*the counterfactual text*) which is the synthesis of a counterfactual argument (*the counterfactual context*). This latter comprises (1) the facts it deals with (the *x* that is *A*); (2) the rules which implicitly it refers to ('There exists a hypothetical rule *R*, which – as far as we know – is valid for all the past, present and future cases in which there is a *x* that is *A*'); (3) the inferential steps which lead from the premises to the conclusions (the steps from 1 to 5, of the above schemes).

On the other hand, there is the judge's applied hermeneutical region composed of nomological knowledge, ontological knowledge and inferential knowledge. Therefore, the judgement of validity results from the comparison occurring between the counterfactual context and the applied hermeneutical region.

At this point, an initial result is achieved: a counterfactual proposition will be correctly judged if the applied hermeneutical region is *adequate.* By *adequate* I mean that:

(1) the facts considered in the counterfactual argument are the same facts belonging to the ontological part of the applied hermeneutical region;
(2) the rules adopted in the counterfactual argument are the same rules contained in the nomological part of the applied hermeneutical region;[31]
(3) the inferences used in the counterfactual argument are the same inferences included in the inferential part of the applied hermeneutical region.

But, owing to the non-eliminability of the ambiguity of many counterfactual arguments, it is not always possible immediately to understand if the applied hermeneutical region is also the adequate hermeneutical region.

Herein lies *the analytical$_h$ problem of counterfactuals; that is, not the problem of determining the truth conditions of a single counterfactual proposition, but the problem of determining the adequacy of the hermeneutical region which the judge must apply to the counterfactual argument synthesized by the counterfactual proposition.*

Let us consider the counterfactual proposition whose latent counterfactual argument I have just expressed in clear and full way: 'If we let a stone drop, it falls to the ground.' In this case, it is quite simple to determine the adequate hermeneutical region to be applied. It must contain ontological knowledge concerning stones and the Earth; nomological knowledge provided by the classical theory of gravity; and inferential knowledge on the usual inferential rules. In such a case, the comparison between the counterfactual context and the hermeneutical region leads us, in an unambiguous way, to judge the validity of the counterfactual.

However, not all cases are so simple and unambiguous. There are some cases in which the ambiguity arises from the ambiguity of the facts involved, others from the ambiguity of the rules *R*, still others from the ambiguity of the inferences used.

Let us exemplify the first two cases. We encounter the first one particularly when we are analysing counterfactuals concerning object *y* and we cannot state with certainty that '*y* is *A*'. Here is a classic example:

'If I struck a match, it would light.'

This is a counterfactual proposition that synthesizes the following counterfactual argument:

(1) The match (*a*) is treated in a particular chemical way (*A*);
(2) There exists a hypothetical chemical rule *R*, according to which every time I strike an object chemically treated in a particular way it lights. As far as we know, this statement is valid for past, present and future cases in which there is an *x* treated in a particular chemical way (*A*);
(3) The lighting of the match (that is, a proposition on $a - P_a$) follows the rule according to which every time I strike an object chemically treated in a particular way it lights (*R*);
(4) It is supposed that an object (*a'*) is an *x* which is chemically treated in a particular way *A*, then *a'* is hypothetically a case contemplated by the rule according to which every time I strike an object chemically treated in a particular way it lights (*R*);
(5) Then the lighting of the match (that is, a proposition on $a' - P_{a'}$) follows the rule according to which every time I strike an object chemically treated in a particular way it lights (*R*).

It would seem that the adequate hermeneutical region to apply is made up of a nomological part with the rule *R*, an ontological part with the knowledge on the usual matches, and the inferential part. But the case is not so simple. Is such a hermeneutical region really adequate? Actually, the ontological context within which the counterfactual argument is developed could be the one considered in the hermeneutical region, but it could equally well be a different one. For example, we could be in the presence of a match treated chemically in such a way to light only when struck on a rough, chemically treated surface. In this case, neither the ontological part of the applied hermeneutical region would be adequate, nor would the entire hermeneutical region. It follows that our judgement of the validity of the counterfactual would be wrong.

Another source of erroneous judgements is provided by ambiguities deriving from the rule *R*. Considering the example of Charles Martel, the hermeneutical region contains a nomological part in which there is a rule such as: 'Those who are not defeated continue to go on.' By means of this kind of nomological knowledge we reach the conclusion that the counterfactual: 'If Charles Martel had not defeated the Arabs, they would have conquered France' is valid. But if our nomological part of the hermeneutical region contained a rule such as: 'Every time an army is not defeated, it stops on the positions conquered', we would make a different judgement.

Reflecting upon these examples, we reach two conclusions: (1) the ambiguities in the factual, nomological and inferential part of the counterfactual context imply difficulties in the choice of the hermeneutical region to apply; (2) different applied hermeneutical regions imply different judgements of validity.

The second conclusion should not astonish us too much. Ironically enough the judgement of validity of a counterfactual, that is, the grasping of a counterfactual, can be likened to the laugh arising from a joke. There are those who do not laugh because the joke is bad, but there are also those who do not laugh because they have not understood the joke, because they do not possess the adequate hermeneutical region. At a certain point in his 1975 *Against Method*, Feyerabend provocatively writes that in Popper's life there was no 'Henriette Taylor'. If someone does not know who Henriette Taylor was this quip falls on deaf ears. The same is true of counterfactuals. *The hermeneutical region has to be adequate to the counterfactual argument expressing in a clear and full way the counterfactual proposition, otherwise the latter cannot be correctly judged.*

However, the fundamental problem of the analytical$_h$ approach that I am presenting reoccurs when we pose the question of how we may identify the adequate hermeneutical region.

The principle of normality and the hermeneutical dialectics

When a hermeneutical region is going to be used, a *principle of normality* comes into play. By *principle of normality* I mean that principle according to which *we apply that particular hermeneutical region which usually we would apply in cases we know (or suppose) to be similar*. In other words, we apply that hermeneutical region we are used to considering to be the most adequate. For example, in the case of the falling stone, it immediately becomes clear that the more adequate region to draw on is the one including the nomological part containing the classic law of gravity; in the case of the match, it is at once clear that the more adequate hermeneutical region is the one having an ontological part containing knowledge about matches which immediately light when struck on a rough, non-chemically treated surface, and so on.

But what happens if the principle of normality leads us to a non-adequate hermeneutical region? Before tackling this question we have to underline that there are two different situations: (1) the judge does not become aware of the non-adequacy of the hermeneutical region he has applied; (2) the judge becomes aware of the non-adequacy of the hermeneutical region he has applied. In the first case, his judgement of validity on the counterfactual is *objectively wrong*, even if it is *subjectively correct*. In other words, for the judge who does not notice the non-adequacy of the applied hermeneutical region, the comparison between the latter and the counterfactual argument ends in a judgement he considers to be correct, even if it is not. *And the game ends here*. The second case is more interesting: the judge notices the non-adequacy of the applied hermeneutical region. Therefore, he modifies its initial hermeneutical region to construct a new one by means of which he re-addresses the counterfactual so as to judge it differently. *And the game can go on indefinitely*.

It should be noted that the awareness of non-adequacy could not only be the result of the hermeneutical collision between the judge's hermeneutical region and the counterfactual argument, but also of the competition between different hermeneutical regions. For, until now, two elements have played the hermeneutical game: the applied hermeneutical region and the counterfactual argument expressing in a clear and full way the counterfactual proposition. But sometimes a third element may come into play: the hermeneutical region of a second judge. In this case, we do not have a *hermeneutical dyad* made up of the judge's hermeneutical region and the counterfactual, but we have a *hermeneutical triad* made up of the two judges' hermeneutical regions and the counterfactual. If, in the first case, we are used to speaking in terms of a *hermeneutical circle*, in the second one, two aspects have to be taken into consideration: (1) the *hermeneutical dialectics*, between the hermeneutical region H applied by the judge A to the counterfactual C, and the hermeneutical region H′ applied by judge A′ to the same counterfactual C; (2) the *two hermeneutical circles*: one between the hermeneutical region H and the counterfactual C, the other one between the hermeneutical region H′ and the same counterfactual C. Now between A and A′ there is a dialectic disputation in which each of the two rivals argues for his own nomological, ontological and inferential knowledge and against the nomological, ontological and inferential knowledge of the hermeneutical region applied by the other.

This is a typical situation that arises especially where there are counterfactuals from social and historical sciences, that is, sciences in which the rules hypothetically connecting the antecedents and the consequents are more disputable than the rules hypothetically connecting the antecedents and the consequents of counterfactuals from natural sciences.

3.3.5 The analytical$_h$ approach

Consequently, a correct awareness of the validity of a counterfactual proposition is not reached by considering it as a single proposition, but as the synthesis of a counterfactual argument. It is exactly such an entire counterfactual argument that must be judged. This is possible either (1) by means of a game with two players, that is, the judge's hermeneutical region and the counterfactual context, or (2) by means of a game with three players, that is, the first judge's hermeneutical region, the second judge's hermeneutical region, and the counterfactual context.[32] This means that to value, as many analytical$_{l-l}$ philosophers have done, the counterfactual proposition separately from the ontological, nomological and inferential context in which it assumes its real meaning, is analogous to considering only the emerged and visible part of an iceberg. However, in such a way, we run the risk of sinking, as many analytical$_{l-l}$ attempts to deal with counterfactuals have sunk. Instead, if we proceed as proposed, this risk should be avoided. Certainly, now *attention has not been paid to the logico-linguistic side of the problem, but to its hermeneutical side.* Consequently, the pivot of the question turns from the truth of a single proposition to the correct way of grasping an entire argument.

Note that if the main feature of the analytical approach is the philosophical precision and exactness, then also the proposed approach has to be considered analytical. And if the *analytical$_{l-l}$ approach* has been meant as an approach characterized, beyond precision and exactness, by the attention to the *logico-linguistic aspect* of the question, we could call the proposed *analytical$_h$ approach*, an approach characterized, beyond precision and exactness, by the attention to the *hermeneutical aspect* of the question. In this way, and beyond a purely nominalistic aspect, I wish to emphasize, firstly, that hermeneutics can be precisely and exactly applied; secondly that its techniques may be used with good results on many occasions; and in particular, that an analytical$_h$ approach may allow us to understand why, and how, the meaning and the validity of counterfactuals can be grasped.

Notes

1. Concepts and objects

1. By *characteristic* of a concept I mean what the Latins called the *nota* of a concept. In Kant and in Frege, we find the German word *Merkmal*. Sometimes *Merkmal* is translated into 'characteristic' sometimes into 'mark'. As just said, I use the first term.
2. The post-predicaments, that is, the properties of categories, concern what is opposed, what is prior, what is posterior, what is simultaneous, the change, and the having. Aristotle discusses them in *Categories*, 10–15. The term 'post-predicaments' was devised by Boetius since Aristotle speaks of them as after the predicaments, that is, after the categories.
3. Of course a conscious use of categories does not necessarily involve any theorization of categories.
4. See Porphyry, *In Aristotelis Categorias.*
5. See Trendelenburg (1846) and Apelt (1891). In antiquity, the logical interpretation had already been supported by Porphyry.
6. For example, Ragnisco (1871) supports it.
7. Let us remember that 'spoken sounds' are 'symbols of affections in the soul' (*De Int.*, 1, 16a 4).
8. Aristotle divides *substance* from the *secondary substances*: 'A *substance* – that which is called a substance most strictly, primarily, and most of all – is that which is neither said of a subject nor in a subject, for example the individual man or the individual horse. The species in which the things primarily called substances are, are called *secondary substances*, as also the genera of these species. For example, the individual man belongs in a species, man, and animal is a genus of the species; so these – both man and animal – are called secondary substances' (*Cat.*, 5, 2a, 10–20). Therefore, from an ontological point of view, we have substance (the real one and the secondary ones) and the accident. From a logical point of view, we have the subject (substance) and the predicates, which can be essential (we are dealing with secondary substances) or accidental (we are dealing with accidents).
9. Aristotle discusses this doctrine in particular in *Topica*, but it will be Porphyry in *Isagoge* who will give it a formal structure.
10. Actually an individual could be predicated of itself: 'Socrates is Socrates'.
11. This is the only point at which I mention the first edition.
12. Remember that 'thoughts without content are empty, intuitions without concepts are blind' (KdrV, p. 93, B75).
13. I will discuss the term 'in general' (*überhaupt*) in the next chapter.
14. Compare, for example, the discussion of causality in P, pp. 53–4, 300–1.
15. Note that I slightly modify the translation indicated in the references, to make it consistent with the other English translation of Kant's writings I use. In particular, I prefer 'knowledge' instead of 'cognition' to translate '*Erkenntnis*', and 'representation' instead of 'presentation' to translate '*Vorstellung*'.

16. With reference to this topic, it is worth reading again §26 of the Transcendental Deduction.
17. With reference to this quotation, it should be noted that the English translators of the *Logik* render '*Inhalt*' with 'intension'. Probably they follow Paton who suggested translating '*Inhalt*' into 'intension' or 'connotation' (Paton, 1936, vol. I, pp. 192–6). A distinction is needed, though. The fact that a concept, considered as partial concept, determines another concept, concerns the *comprehension* of the latter, that is, its content of characteristics. Moreover, the fact that objects fall under one concept concerns the extension of that concept. It seems to me, therefore, that in neither case speaking of 'intension' is correct.
18. Because of the many ways in which the Fregean terms '*Sinn*' and '*Bedeutung*' have been translated, in the following I will maintain them. Therefore, the English translations of the passages containing them will be slightly modified.
19. Actually, this way of speaking should be considered wrong, according to Frege. On the one hand, we should speak of identity between the two expressions denoting the two functions, and, on the other hand, it is not the functions that are said to be identical, but their graphs. Hereafter, at least in non-ambiguous cases, I will not specify whether I am referring to the sign of the function, rather than to the function, or to the sign of the argument, rather than to the argument.
20. Actually, as emphasized in the previous note, this claim is not correct. According to Frege, not two functions, but two graphs can be considered identical: 'the possibility of regarding the equality holding generally between values of functions as a [particular] equality, viz. an equality between graphs is, I think, indemonstrable; it must be taken to be a fundamental law of logic' (Frege, 1892a, p. 26). At this point, it is worth recalling a wonderful *escamotage* by means of which Frege makes his argument consistent. As seen, a function is an unsaturated expression needing be saturated by something heterogeneous to it, namely an argument, an object. Nevertheless, there are type-I-functions that can be a problem, namely constant functions such as $f(x) = 2$, which, at a first glance, do not have any blank to be saturated. How should they be treated? 'There are functions' – Frege writes – 'such as $2+x-x$, or $2+0x$, whose value is always the same, whatever the argument; we have $2=2+x-x$ and $2=2+0x$. Now if we counted the argument as belonging with the function, we should hold that the number 2 is this function. But this is wrong. Even though here the value of the function is always 2, the function itself must nevertheless be distinguished from 2; for the expression for a function must always show one or more places that are intended to be filled up with the sign of the argument' (Ibid., p. 25). Thanks to this philosophical manoeuvre, any constant function can be considered as containing one or infinite blanks, which, though algebraically eliminating each other, allow us to interpret it as an unsaturated expression.
21. Compare 'Compound Thoughts' (Frege, 1923–6) and other unpublished essays, now edited in *Posthumous Writings*.
22. With reference to the term 'image', compare what I said in the Introduction.
23. To be honest, there is also a passage which could be understood as claiming the priority of language over thought: 'Kerry holds that no logical rules can

be based on linguistic distinctions; but my own way of doing this is something that nobody can avoid who lays down such rules at all; for we cannot come to understanding with one another apart from language, and so in the end we must always rely on the other people's understanding words, inflexions, and proposition-construction in essentially the same way as ourselves' (Frege, 1892b, p. 45). Certainly, here Frege points out the importance of language, and certainly he is claiming the priority of language. Nevertheless, he is not speaking about the priority of language in comparison with the thought, but about the priority of the perfect language he is considering and he wants to construct. And the perfect language is structurally isomorphic to the thought.

24. Actually in the English translation of Frege's writings I am working with, the term 'mark' is used instead of the term 'characteristic'. But, as said, I adopt this latter locution. Consequently, I modify the translation of the Frege passages I quote.
25. Several times, Frege emphasizes both the distinction between 'identity' and 'having the same extension', and the fact that a concept and its extension are not the same. For example, 'if we write $x^2 - 4x = x(x-4)$ we have not put one function equal to the other, but only the values of one equal to those of the other' (Frege, 1891, p. 26); or 'if he [Kerry] thinks that I have identified concept and extension of concept, he is mistaken' (Frege, 1892b, p. 48).
26. Frege had necessarily to write his essays using natural language, and so he had to accept all its ambiguities and traps. He was well aware of this fact and, whenever he could, he underlined such ambiguities and traps by using the perfect language (compare Frege, 1892b, p. 47, fn. †).
27. Compare my discussion of fictive objects in the third chapter.
28. Furthermore, this assumption shows once more that according to Frege language is not prior. What is prior is the thought, in particular the true thought, and then what makes it true, namely the structure of the existing objects. Also according to Aristotle non-referring names were possible, but these disappear as soon as we deal with true *propositions* (the only ones describing being). There is a difference, however. For Aristotle, it is ontology and its logical structure which are prior and then there are thought and the mental contents, which are the faithful mirror (then reproduced by language) of whatever exists. For Frege, it is the thought with its logical structure (then reproduced by the perfect language) which is prior and then there is ontology, which is its faithful mirror.
29. With reference to the notion of vagueness, compare Boniolo and Valentini (2007).
30. The original German title is much more meaningful: *Substanzbegriff und Funktionbegriff*, namely *Concept as substance and concept as function*.
31. It should be noted that a metaphysical realist can argue for the reduction of mental acts to brain processes. However, this possibility does not have any importance for our discussion.
32. With reference to the two terms '*ectypus*' and '*archetypus*', see Kant (1986, pp. 99–106); also the next chapter.
33. There are those who maintain that Pegasus is observable since it can be observed that it does not exist. Here, we have a different definition of 'observable'. From what I have asserted, 'observable' implies intersubjective

perceivability detectable by all those who wish it. Actually I could affirm that Pegasus is hypothetically observable (in well-determined circumstances), but as soon as this hypotheticity vanishes, its hypothetical observability (in the same determined circumstances) also vanishes. That is, I could claim that Pegasus is an intentional object that is also conjecturally empirical, and therefore observable. But, as soon as I have detected that there exist no aspects of metaphysical reality which can be constituted by those particular intentional properties, then Pegasus becomes non-observable. Note that the history of science is full of cases of this kind of null experiment. For example, Michelson and Morley's experiment to detect absolute motion was a null experiment.

34. Notwithstanding this, introducing this type of field raises new problems such as (1) explaining the reason why interaction is not universal; (2) understanding the origin of the mass of Higgs boson; (3) solving the question of the interaction between the field and the graviton (responsible for the gravitational interaction) in a less ad hoc way (one gets around the great curvature which the universe is said to have due to the field, by considering vacuum with a negative curvature, so that, by adding algebraically to this negative curvature the positive curvature due to Higgs field, space becomes flat).

2. Laws of nature

1. Kant writes: 'I freely admit that the resemblance of David Hume was the very thing that many years ago first interrupted my dogmatic slumber and gave a completely new direction to my research in the field of the speculative philosophy' (P, p. 10, 260).
2. This is my neologism. Some authors prefer to speak about *lawhood* (compare Vallentyne, 1988).
3. Note that 'nature' in this context must be interpreted in a Kantian way, as discussed below in the section 'Nature *überhaupt* and experience *überhaupt*'.
4. Now 'nature' must be interpreted in the neopositivist way.
5. Here a third meaning of 'nature' is to be understood, that is, in a metaphysical way.
6. With regard to this aspect, it should be noted that many interpretations of Kant's doctrine of laws of nature are vitiated precisely by this misunderstanding between lawness and lawfulness.
7. We should observe that, for Schlick, speaking about causality is speaking about law: 'in science the [real] dependence [among the events] is always expressed in each case by a *law*; thus causality is just another word for the existence of law' (Schlick, 1931, p. 177).
8. 'The confirmation of a prediction never ultimately *proves* the presence of causality' (Ibid., p. 187). That is, Schlick was well aware of the impossibility of inferring the truth of a universal statement from a single confirmation, or even from many confirmations.
9. In 1931 Schlick cannot know the following critiques by Popper (1934) and Carnap (1936) on the impossibility of conclusively verifying singular statements.
10. Schlick writes explicitly about the possibility that there are regularities due to chance (Ibid., p. 184).

11. Later, in 1935 in his 'Are Natural Laws Conventions?', Schlick will observe that his position does not involve a conventionalist interpretation of laws but a particular kind of realism towards them.
12. Popper proposed both abandoning the notion of 'significance criterion' and working on the notion of 'demarcation criterion', and abandoning the concept of 'verification' and using the concept of 'falsification'. Schlick suggested a solution rooted in the so-called second Wittgenstein. Carnap, finally, argued that from the notion of 'verification' we should move to the notion of 'confirmation'.
13. I call 'classical regularists' the authors who preceded the modal turn of the 1960s, and 'new regularists' those who worked after then.
14. In 1976 a new edition, edited by Salmon was published. It was titled *Laws, Modalities and Counterfactuals* (University of California Press, Berkeley). Salmon's foreword was then published again in *Synthese* (see Salmon, 1977).
15. On the notion of 'explication', see Boniolo (2003).
16. Beyond Nagel, it seems that only Salmon (1977) and Jobe (1977) have devoted a detailed analysis to this work.
17. Actually he inserts another four requirements, but we may leave them aside, since they concern mainly statements different from laws of nature.
18. Let us note that if our philosophy starts from physics, problems arise as soon as we move to biology. This was a big problem for the physically-modelled philosophy of science before the 1960s.
19. 'Since physical possibility is a category to be defined in terms of nomological statements, it would be circular to use, in the definition of such statements, this category' (Reichenbach, 1954, p. 12). Note, as I will show later, that the new metaphysicians try to break the circularity by ontologizing one of the terms: either they define nomologicity (given at the epistemological level), by resorting to modality (given at the ontological level), or they define modality at the epistemological level, by resorting to nomologicity (given at the ontological level).
20. I can really measure the temperature of the room, and then I know exactly its value (we are in the case of the actual application of P_0). However, I could not measure the temperature, but if I did I would know it (we are in the case of the possible application of P_0).
21. Reichenbach connects this requirement with his theory of probability, as he shows in the Appendix of his 1954 work analysed above.
22. With reference to attempts to explicate the notion of confirmation and their failures, see Boniolo and Vidali (1999, ch. 5).
23. Nagel, as Goodman (1954) had already done, suggested modifying the original Humean approach by freeing it from psychologism (Nagel, 1961, p. 62). Nevertheless, unlike Goodman, he did not suggest any good alternative. Note, moreover, that the Humean psychologism allowed him an account of the reason why the induction worked: a problem not tackled by Nagel.
24. This is the text of a lecture delivered at the New York Philosophical Circle on 11 May 1946.
25. Goodman proposed his view in three lectures given at the University of London on 21, 26 and 28 May 1953. They are contained in Goodman (1954–73, ch. II).

26. Actually Goodman did not focus his attention on representations, but on predicates and hypotheses. His was a different way of considering the same things, but less relevant to what I am discussing. In my view a hypothesis is a representation; compare next chapter.
27. Note that there is a difference between the non-projectable hypothesis (supported, inviolate, unexhausted but not overridden by a conflicting hypothesis) and the unprojectable hypothesis (unsupported, violated, exhausted, or overridden by a conflicting hypothesis).
28. Beyond Lewis's line of thought (1973a, 1973b, 1983) see Kripke (1963), Stalnaker (1968) and Stalnaker and Thomason (1970). Along the necessitarist line, see Sellars (1948), Pargetter (1984), McCall (1984) and Vallentyne (1988). With reference to the *de re* approach, see Tooley (1977), Dretske (1977) and Armstrong (1978), and their ancestors Kneale (1950 and 1961) and Molnar (1969).
29. Actually, as Armstrong himself recognizes, not all the regularists identify *tout court* laws of nature with Humean uniformities, since the former are a subclass of the latter (see Armstrong, 1983, pp. 13–14). It should be noted that the main attempt of the standard view was exactly that of defining which Humean uniformities could be considered as laws of nature.
30. Recall that Reichenbach excluded that the statements with empty antecedent could be considered laws; instead Nagel is more cautious, since he knows there are statements that can be vacuously true.
31. Another example is given by the following statement 'All the crows are black'. For the regularist, this expresses a law affirming the non-existence of non-black crows. Nevertheless, as should be evident, the existence of such non-black crows cannot be excluded, even if no non-black crow has been observed until now. If crows lived in regions with perennial snow – Kneale suggested – evolutionary biology could give good reasons to believe in the possibility they could have white-feathered descendants. Therefore the statement above cannot be considered as a law (Kneale, 1950, pp. 122–3). A similar example is discussed by Popper (see Popper, 1959, pp. 483–4).
32. For a critique of the reading of Kant with an Humean interpretative grid, proposed by contemporary commentators in English, see Lee (1981, p. 407); for a criticism of those who interpret Kant to accommodate him to the results of twentieth-century physics, see Friedman (1994, pp. 27–8). Many scholars underestimate Kantian architectonics, sometimes reducing the problem of laws, for example, mostly to the *Metaphysical Foundations of Natural Science* (see Friedman, 1992a, 1992b and 1994) and thus neglecting the *Critique of Judgement* (see Friedman, 1994; O'Shea, 1997). As far as the inductivist reading of Kant is concerned, I will be more specific further on. However, there are also those who read Kant from the point of view of their own philosophy of science, as explicitly confessed by Kitcher (1994, p. 270, fn. 1). With reference to this aspect, I share Kitcher's notion of utilizing Kant to propose a different philosophy of science, and of reading Kant by using the philosophy of science. However, I would be suspicious of the possibility of forcing Kant to affirm what we believe valid in the contemporary philosophy of science or epistemology.

33. Precisely three times in the *Kritik der Urteilskraft* (once in A318 and twice in A404); twice in *Die Religion innerhalb der Grenzen der blossen Vernunft* (A3 note and A123 note); and once in the *Anthropologie in pragmatischer Hinsicht* (A171). I am quoting from Kant's *Werkausgabe.*
34. Surprisingly enough, this distinction is not adequately considered. For instance, Buchdahl writes 'lawlikeness is used (by myself) in the sense of conformity to law (*Gesetzmässigkeit*)' (Buchdahl, 1965, p. 25). Here Buchdahl refers, as can be understood from the corresponding context, to conformity to a law of nature, that is, to its *lawfulness.* He refers neither to the conformity to a law of a statement, that is, to its *lawness;* nor to the fact that some statements have the form of a law but are not said to be laws, that is, to their *lawlikeness.* Nevertheless, there are English commentators, such as Beck, who translate *Gesetzmässigkeit* correctly by 'lawfulness'.
35. Of course here 'deduction' is intended in Kant's juridical sense: 'Jurists, when speaking of rights and claims, distinguish in a legal action the question of right (*quid iuris*) from the question of fact (*quid facti*); and they demand that both be proved. Proof of the former, which has to state the right or the legal claim, they entitle the *deduction*' (KdrV, p. 120, B116).
36. However, reason can fall into error in two other ways: (1) it can make a logical error; (2) it can make a methodological error. In the first case, the correct deductive rules of the syllogistic inferences are not applied. That is, we have 'deficiency in judgment [that is . . .] just we have what is ordinarily called stupidity, and for such a failing there is no remedy. An obtuse and narrow minded person to whom nothing is wanting save a proper degree of understanding and the concepts appropriate thereto, may indeed be trained through study, even to the extent of becoming learned. But such people are commonly still lacking in judgment (*secunda Petri*), it is not unusual to meet learned men who in the application of their scientific knowledge betray that original want, which can never be made good' (KdrV, p. 178, B172–3, fn a). In the second case, which I will discuss later, the correct judgement under which the given empirical situation has to be subsumed is not applied. Note that Kant by reasoning *secunda Petri* means reasoning correctly from the logical point of view, that is, as it is indicated in the textbook of logic, *Dialecticae Institutiones*, written by Petrus Ramus in 1543; for an English translation partially modified, see Ramus (1574).
37. See Kant (1986, pp. 99–106). Note that there is also another possibility: Leibniz's pre-established harmony, which Kant also rejects (see KdrV, p. 174, B167).
38. 'We should then be proceeding precisely on the line of Copernicus' (KdrV, pp. 22–3, BXVII).
39. Correctly Paton calls this way of arguing the 'transcendental proof' (Paton, 1936, vol. II, pp. 103–6; compare Bayne, 2000, pp. 214–15).
40. Buchdahl (1971) differentiates between 'nature' and 'order of nature', and between 'experience' and 'order of experience'. He puts the first element of the two pairs into what I have called the metaphysical level and the second element into what I have called the empirical level. But this proposal does not work so well. For already *natura formaliter spectata* is an ordered type of nature, even if in the sense of its possibility in general. This naturally applies also to experience if considered as experience 'in general'.

41. It is important to be aware that 'though all our knowledge begins *with [mit]* experience, it does not follow that it all arise *out of [aus]* experience' (KdrV, p. 41, B1, my emphasis).
42. To prove this two-fold meaning there is the following passage from the *Prolegomena* (numbers (1) and (2) in parenthethes indicate the first and second meanings): 'But how does this proposition: that judgments of experience(1) are supposed to contain necessity in the synthesis of perceptions, square with my propositions, urged many times above: that experience(1), as *a posteriori* cognition, can provide merely contingent judgments? If I say: experience(1) teaches me something, I always mean only the perception that is in it – for example, that upon illumination of the stone by the sun, warmth always follows – and hence the proposition from experience(1) is, so far, always contingent. That this warming follows necessarily from illumination by the sun is indeed contained in the judgment of experience(2) (in virtue of the concept of cause), but I do not learn it from experience(1); rather, conversely, experience(2) is first generated through this addition of a concept of the understanding (of cause) to the perception' (P, p. 58, footnote *, 305) For the second meaning: 'even, therefore, with the aid of [pure] intuition, the categories do not afford us any knowledge of the things; they do so only through their possible application to *empirical intuition*. In other words, they serve only for the possibility of *empirical knowledge*; and such knowledge is what we entitle experience' (KdrV, p. 162, B147).
43. It should be pointed out that we are dealing with the 'metaphysics *of* nature' and not with 'metaphysics *in* nature'.
44. From the passages mentioned it follows that Kant's view of 'physics' does not have exactly the same meaning it has today. For, according to Kant, 'physics' is a discipline characterized by both its logical structure, its subject matter, and its having a philosophical part, especially a metaphysical one. Kant is continuing the tradition which interprets physics as mathematized natural philosophy. This a tradition which was epitomized by Newton with his *Philosophiae Naturalis Principia Mathematica*, but that had as forerunners in astronomy Copernicus (with his 1953 *De revolutionibus orbium caelestium*), and, of course, Galileo Galilei (with his 1632 *Dialogo sopra i due massimi sistemi del mondo*). With reference to this point, see Boniolo (2004). It is also worth mentioning that in the introduction to the *Metaphysical Foundations*, Kant speaks explicitly about 'philosophers of nature' and 'mathematical physicists' (M, p. 8; 472). In Kant's opinion (as for Newton) the true philosophers of nature are the mathematical physicists, that is, the physicists both dealing, in a systematic way, with the a priori principles governing their field, and using mathematics. A last remark should be made to point out how the notion of 'philosophy of nature' changed in the nineteenth century as a result of romantic German idealism.
45. This is an extremely interesting point since Kant, precisely on the grounds of mathematics, succeeds not only in making coherent the formulation and the use of the *Metaphysical Foundations*, but also because he succeeds in giving a brilliant answer to a problem which has troubled many philosophers of science: why is mathematics so successful in the physical sciences? On this topic, see Boniolo and Budinich (2005).

46. Note that they are three laws of Newtonian mechanics, but they are not 'the' three laws of Newtonian mechanics!
47. To grasp this point is sufficient to give a brief comparative look at the pure principles and at the corresponding first principles. If this interpretation is valid, then the position claimed by Friedman is only partially correct. Friedman proposes that 'Kant derives these Newtonian laws of motion (the laws of dynamics) with an *a priori* proof on the basis of the corresponding transcendental principles of the understanding (the analogies of experience) replacing in them the empirical concept of matter' (Friedman, 1994, p. 33). Even if this were possible (as I think it is) Kant does not do it. He finds, as we have seen, the three principles of mechanics by reflecting on the definition of the concept of matter from the point of view of the categories of relation. Only a posteriori is it possible to see that we could also have obtained them on the basis of the analogies with their particularizations on the grounds of the concept of matter (and not simply implementing that concept in them). And this, I repeat, could be possible only for the principles of mechanics, not for those of phoronomy, not for those of dynamics, and not for those of phenomenology.
48. Empirical lawfulness will be analysed later.
49. This difference between judgements of perception and judgements of experience appears in the *Prolegomena* and in the *Logic* (L, § 40), while in the *Critique of Pure Reason* only the judgements of experience, called simply empirical judgements, are mentioned. Moreover, it is still in the *Prolegomena*, and not in the *Critique of Pure Reason*, that is possible to find the famous difference between *synthetic a priori* and *synthetic a posteriori* (P, pp. 17–18, 267–9). I recall that, in Kant's view, perception 'is empirical consciousness, that is, a consciousness in which sensation is to be found' (KdrV, p. 201, B207).
50. It is worth mentioning that in the quoted passage of the *Prolegomena* the qualification 'in general' still appears. That Kant has replaced the expression 'consciousness in general' with 'transcendental unity of apperception' has not prevented rivers of ink flowing on this topic, as noted by de Vleeschauwer referring to Amrhein's work (*Kants Lehre v. Bewusstsein überhaupt*, Berlin, 1909). De Vleeschauwer also points out that the expression 'consciousness in general' does not appear in the first edition of the *Critique of Pure Reason*, although it appears four times in the *Prolegomena*, and only appears once in the second edition (see de Vleeschauwer, vol. I, 1936, pp. 464–5). Regarding the division between judgements of experience and judgements of perception, Buchdahl (1965, p. 192, n. 1) emphasizes that it has been denied by many commentators, including Cassirer and Kemp Smith.
51. In §19 there is also an answer to the question raised in the *Metaphysical Foundations* on the necessity of a new definition of judgement (cf. M, p. 11, 475, fn).
52. It should be noted that it is not true that Frege was the first (as is claimed by some historians of logic) to propose an alternative theory to the Aristotelian-medieval one according to which 's is p'. In fact, as we can see, Kant had already proposed an alternative theory, even if it was obviously different from that proposed by Frege a century later.

53. Kant can be seen as the one of the first philosophers who explicitly and exhaustively proposed the theory-ladenness of the empirical data; a thesis proposed later by many philosophers of the so-called 'new philosophy of science' tradition as if nothing similar had ever happened before.
54. In the recent literature on this issue, we find (1) commentators such as Strawson, who, following Lovejoy, accuse Kant of committing a '*non sequitur* of numbing grossness' in passing from the necessity of causality at the transcendental level to the necessity of causality at the empirical level (Strawson, 1966, p. 137; compare Lovejoy, 1906 and the critical discussion by Buchdahl, 1969; Beck, 1976; Allison, 1983); (2) commentators such as Buchdahl (1965 and 1971), who claim that there is no relation, in particular any logico-deductive relation, between the second analogy and the empirical laws, and that to grasp the question clearly induction must be considered (this is the so-called *weak solution*, apparently shared also by Bayne, 1994); (3) commentators such as Friedman (1992a, 1992b and 1994; criticized by Allison, 1994), who hold, contrary to Buchdahl, that there is a relation, made possible thanks to the metaphysical principles and, therefore, that all the problems can be solved only by resorting to the understanding (the so-called *strong solution*); (4) commentators who defend the weak solution, but try to tackle the difference by reading the pure principles of the understanding as syntactical rules, while the nomological validity implies semantic questions (see Butts, 1969 and 1971); (5) commentators who defend the strong solution and consider the pure principles of the understanding as conceptual rules of the second order, which strictly imply the empirical laws, considered as conceptual rules of the first order (see O'Shea, 1997); (6) commentators seeking an intermediate solution (see Guyer, 1990a); (7) commentators who do not tackle the problem directly (see Kitcher, 1994). I do not want to confront these positions individually, although it will be clear that I do not accept the Lovejoy–Strawson position, while I am sympathetic both towards Buchdahl (while nevertheless rejecting any form of inductivism), and with Bayne. Regarding the various positions, see Thöle (1991).
55. Note that the analogies of experience, which are regulative, constitute the lawfulness of nature. As we are going to see further on and as Kant himself specifies (KdrV, pp. 545–6, B691), the analogies *regulate* in the sense that they impose laws, and thus *constitute* nature as *natura formaliter spectata*. This demands a further remark. There are two meanings of 'regulative' to be distinguished: (1) 'regulative' in the sense of the pure laws (rules) which are constitutively imposed to rule nature; and (2) 'regulative' in the sense of heuristic rules helping scientific research.
56. This was an objection raised, for example, by Schopenhauer (1847, §23).
57. It is not trivial to claim that Kant had correctly interpreted Hume's challenge, and that the second analogy of experience is in that respect a good answer. On this issue, apart from the references already mentioned, see Beck's (1978) classical work, Thöle (1991), Bayne (2000) and the references therein.
58. I derive this reflection from Lee (1981, pp. 406–7), who, however, considers only physical contingency/necessity and logical contingency/necessity. However, we should also trace a distinction between formal logic and transcendental logic and, thus, consider also the transcendental

contingency/necessity. Without the latter, it is not possible to correctly grasp the principle of the second analogy.

59. For instance, Kitcher (1994, p. 257), and O'Shea (1997, pp. 242–8) support the first thesis; Scaravelli (1968, pp. 369–371), and Guyer (1990b) the second.
60. It should be noted here that 'unity' is not intended in the transcendental sense, as it was the unity allowed us by the 'I think'. Instead now it must be intended as a methodological unity.
61. For Scaravelli, the first multiplicity is empirical; the second belongs to the determinations of space and time.
62. Note that the English translation renders '*überhaupt*' by 'as such'.
63. With reference to this point Kant, in the *Metaphysical Foundations of Natural Science*, writes that 'the word nature already carries with it the concept of laws, and the latter carries with it the concept of necessity of all determinations of a thing belonging to its existence' (M, p. 4, 469).
64. From these three principles of reason, Kant derives also that according to which 'non datur *vacuum formarum*', and its correlate '*datur continuum formarum*'.
65. The use of the term 'regulative' in this case is very different from the use of the same term whenever the maxims of reason are concerned. In this case it is intended as something which necessarily rules the relation among phenomena in order to have nature 'in general'. Therefore it is intended in a constitutive sense. Instead now 'regulative' concerns the regulation of scientific research, guiding it, and the construction of the unity. In brief, in the former case, 'regulative' was used epistemologically to make lawfulness possible; in the latter case, it is used heuristically in order to make systematic unity possible. Note that Kant himself disambiguates this twofold possibility (cf. KdrV, pp. 545–6, B691).
66. A different way of arguing against the alleged Kantian contradiction as far as the maxims are concerned can be found in Kitcher (1998). In this paper, Kitcher arrives at the same conclusion that I reach but by focusing on the system of unity.
67. As far as their being *as-ifs* is concerned, it has to be noted that it is from Kant's theory of ideas, always seen in the form of *as if*, that Vaihinger derives his fictionalism, which will be discussed in the next chapter (see Vaihinger, 1911, pp. 243–81). Vaihinger's interpretation is not shared by Adickes (1927), who, instead, suggests distinguishing a fictive from a hypothetical use of *as-if*. Nevertheless, Vaihinger differentiates fictions from hypotheses, considering only the first ones as *as-if*. The hypotheses cannot be such: they have a theoretical value, while the fictions have only practical value. Analogously, both the maxims of reason, which we have already met, and the maxims of the capacity of reflective judgement, which will be dealt with later, have only practical and not theoretical (not cognitive) value.
68. Regarding this point, it should be stressed that Kant uses the term 'reflection' exactly in the same way as the term '*reflektierende*', characterizing, in the third *Critique*, one of the two ways according to which the capacity of judgement works.
69. To be precise, these two expressions do not belong to the transcendental deduction of the maxims of reason; they are introduced immediately afterwards (KdrV, pp. 553–4, B704–5).

70. There are also commentators, for example Lee (1981, pp. 405–8) who seem to be immune to such inductivist contamination. With reference to Lee's position, it should be said that it is based more on what was proposed by the 'new philosophy of science' than on Kant.
71. Here it should be recalled that in the Appendix Kant speaks about the leading thread of reflection. On several occasions, Kant thinks of maxims as a heuristic device (for example, KdU, pp. 24–5, 185).
72. If my interpretation is correct, then there could be problems in not applying correctly the reflecting capacity of judgement, since this would lead to hypothesizing the wrong universal which makes that particular possible. Moreover, it should be recalled that this error has nothing to do with the *secunda Petri* error. As said, this latter is an error due to an infringement of the correct deductive inferences, the former concerns the incorrect way of giving cognitive significance to a particular. Note that Kant speaks also of the methodological role of experience in checking, and possibly rejecting, the empirical law proposed. That is, Kant is well aware of the difference between what we now call 'confirmation' and 'falsification' of the empirical laws (KdrV, pp. 20–2, BXII–XVII and pp. 297–8, B349–52; P, pp. 42–3, 290–1). On this issue, see Scaravelli (1968, pp. 403–9) and Lee (1981, p. 413).
73. For instance, KdU, pp. 236–7, 360–1; pp. 251–2, 372–3; pp. 257, 377; pp. 268, 388–9; pp. 287–8, 404; p. 294, 409–10; § 78, 410–15 passim; pp. 302, 418. Remember that the Appendix has also pointed out the complementarity between causality and purposiveness from the point of view of an analysis of scientific knowledge.
74. Kitcher (1998) arrives at the same result by emphasizing the role of the system. Rescher (1970), instead, emphasizes the role of the legislating understanding. Note that both of them speak in terms of lawfulness, which as we have seen is not the right term
75. As is known, this is the suggestion of many philosophers of biology, who, starting from the regularist presupposition that laws are unrestricted statements, deny the status of law to biological generalizations; this position is supported by, for example, Matson (1958), Smart (1963) and Beatty (1995).
76. Note that the emphasis on the system, without referring to Kant, has had a great role in contemporary philosophy of science, since Duhem (1914) and Poincaré (1902–1907). More than that, some scholars have invoked the system as a possible way out for both the general problem of lawness (for example Nagel, 1961) and the particular problem of biological lawness (for example, Ruse, 1970; Waters, 1998). It is worth mentioning also Kitcher (1994 and 1998), where an analysis of the Kantian notion of 'system' and an attempt to move it into the contemporary philosophy of science can be found.
77. It is worth recalling that the ancient Greeks did not use the term *nomos* with reference to laws of nature, but to laws among men.
78. With reference to the role played by the epistemological relation *R* in the explanatory context, see Boniolo (2005).
79. Of course, here Duhem's lesson contained in *The Aim and Structure of Physical Theory* (1914) should be recalled.
80. The same aversion to the isomorphic and mirror-like theories of representation is shared by Suárez (2003 and 2004).

3. Theories, models, thought experiments and counterfactuals

1. Observe that in the English translation the term *'Bild'* has been given by 'picture'. Recall what was said in the Introduction.
2. It should be noted that Maxwell had already proposed a similar interpretation.
3. Of course here the question of the underdetermination of theories by data is on the stage, too.
4. Let us skip here the question concerning what, for Hertz, 'the laws of thought' are.
5. It is worth noting that this passage was published in 1894, that is, 40 years before the German edition of Popper's *Logic of Scientific Discovery* (1954), where the thesis of falsifiability and the asymmetry between verification and falsification were presented with great emphasis. However, the awareness of the asymmetry between verification and falsification was widespread among the pre-World War II European philosophers of science.
6. Actually, the two physicist-philosophers differ in relation to the epistemological status of the principles supporting the theories. In particular, according to Hertz, they are a priori but distinct from 'the laws of thought', while according to Boltzmann, they are 'the laws of thought' and even if they appear to be not empirically refutable, actually they are. They are empirically testable and not yet falsified propositions which have succeeded in cultural evolution.
7. Note that, according to Duhem (1914), the theory is a model in the sense that it is an axiomatized system connected to the empirical world by means of correspondence laws. As is well known, this is also the core of the interpretation proposed by Carnap, Hempel, Nagel, and Braithwaite. More recently, van Fraassen (1980 and 2002) and Rivadulla (2004) have expressed, *mutatis mutandis*, similar views.
8. As far as I know, among the contemporary philosophers only Fine (1993) seems to be aware of the importance of Vaihinger's proposal.
9. What I call the 'focusing model' shares many characteristics with what is also called a 'mediating model', that is, a model mediating between a 'governing theory' and the phenomena to which it is applied; see Morton (1993) and Morrison and Morgan (1999).
10. Recall the difference (compare Chapter 1) between *abstrahere aliquid*, that is, to abstract what is common (which in our case must be interpreted as 'to abstract what is interesting'), from *abstrahere ab aliquo*, that is, abstracting from what is not common (which in our case must be interpreted as 'to abstract from what is not interesting').
11. It is sufficient to think that a mole of gas contains 6×10^{23} (Avogadro's number) particles. This means that in order to find the equation of the motion for such a system, 1.8×10^{24} equations should be solved by means of 3.6×10^{24} initial conditions. Even if we could solve this enormous system, we should sum the 3.6×10^{24} solutions to obtain the value of the macroscopic variables in which we are interested.
12. Of course, the matter changes when an entire theory is interpreted instrumentalistically.

13. Actually there are some problems with the predictive power of the Lotka-Volterra model. With reference to this point and, in general, to the mathematical models, see Israel (2003).
14. If a phenomenological model is constructed only to save the phenomena, it means that it correlates the facts but it cannot explain them. At most we have, as Cartwright (1983) correctly proposes, a 'simulacrum account of explanation'. That is, we have something which has the form of the explanation but not its substance. Actually, each class of models allows us a 'simulacrum account of explanation'. However, each class offers a different kind of 'simulacrum', and this fact should be taken into consideration.
15. The content of this section summarizes Boniolo et al. (2002).
16. Of course, the NTB is neither completely independent of or separable from the set of the experimental data; some assumptions are grounded on empirical values (for example, the independence of the nuclear interaction from the charge is grounded on empirical spectroscopic data).
17. Redhead (1975 and 1980) points out that the new 'dress' can be larger than the older one. In this case, there is what he calls a 'surplus' structure in the mathematical formulation of a physical theory.
18. A more or less analogous analysis can be made of the so-called 'synergetics', proposed by Haken (1978).
19. The idea that thought experiments are arguments is defended, even if inside a different philosophical context, especially by Norton (1991 and 1996). Note that Popper (1959, Appendix *XI) also indicated this aspect.
20. In a passage contained only in the fourth edition of Mach (1883, ch. 2, § 6.10).
21. Actually, Newton considers this case, too; but he achieves a contrary result. For he discusses the two-sphere experiment in two steps. In the first, the two spheres are the only two bodies in an endless vacuum. In the second, Newton inserts other bodies and concludes that if the wire tension were observed and it were the same as before, it would be necessary to infer that the two spheres move and that the other bodies are at rest. From a historical point of view, it is interesting to note that Mach, in the cited passage above, claims that 'this experiment is unrealizable (and therefore) this idea has no significance'. It would seem that Mach, starting from the empirical impotence of the thought experiments, wants to deny their conceptual value. Yet this empirical impotence is exactly what characterizes thought experiments, as Mach himself admits (1905, Ch. 11), where he gives, paradoxically enough, the first complete theorization of thought experiments. Some historical remarks are worth making here. (1) In fact, Mach's first discussion of thought experiments is in Mach (1896–97), later rewritten as ch. 11 of Mach (1905). (2) Nevertheless the first analysis of the topic, and perhaps the creation of the term *Gedankenexperiment,* must be attributed to Örsted, who discussed it within a Kantian framework (see Witt-Hansen, 1976). (3) Mach, however, was the first who defended the thesis according to which the thought experiment is a kind of precursor of real experimentation. This thesis was criticized, for example, by Meinong (1907, § 15) on the basis of the autonomy of thought experiments from reality. (4) This debate, even if in different language and with a different philosophical source, has been revitalized more recently. On one side of the barricade there are

scholars such as Sorensen (1992) and Gooding (1990 and 1993) supporting the idea that there is a strong continuity, if not a superposition, between thought experiments and real experiments; on the other side, there is Brown (1991 and 1994) arguing for an extremely idiosyncratic Platonic idea on the autonomy of thought experiments. (5) The history of philosophy is full of rhetorical thought experiments. For example, recall the thought experiment of the blind man who suddenly acquires his sight, proposed by Berkeley (1709). Before Berkeley, it was discussed also by Locke and Molineaux. After Berkeley, it was reanalysed by Voltaire and de La Mettrie. Other remarkable examples are given by the thought experiment of the marble statue, proposed by Bonnot de Condillac (1754) to argue for his sensistic conception and the thought experiment of the young man imprisoned in a hole in the earth, argued by Arnobius the Older in the fourth century against the Platonic theory of reminiscence and used by de La Mettrie (1745) both to support his sensism and to criticize Cartesian nativism. In the present age, this tendency to use rhetorical thought experiments, which had its climax in the scholastic age and in the seventeenth century during the epistemological debate, has gathered new force within some of the analytical ways of philosophizing. Criticisms of the use of thought experiments in parts of the analytical tradition can be found in Massey (1991) and Gale (1991).

22. This is also stressed by Norton (1991).
23. The Machian thought experiment could also be considered fallacious if we used the criterion for a good thought experiment proposed by Popper (1959) in his Appendix *XI. Popper, as we have said, emphasizes the argumentative power of some thought experiments, which he calls 'imaginary experiments'. He affirms that an argumentative admissible thought experiment must start from positions accepted by the supporters of the opposed thesis. If his criterion were adopted we should eliminate a great part of the thought experiments from the history of sciences and philosophy. Koyré (1960) explicitly replied to Popper, but his work is also interesting because it contains a psychological explanation of the genesis of a thought experiment quite different from that proposed by Mach (1905).
24. Bohr proposes a slightly modified fictional mechanism. The box is hung to a balance by means of which we measure its mass with an arbitrary precision Δm. But this enables an approximation of Δq of the index of the balance and thus an approximation of Δp of its momentum, correlated to Δq by the first Heisenberg relation. This uncertainty of the box momentum has to be less than the total momentum due to the gravitational field acting, for a time T equal to the length of the entire process, on a body with mass Δm. It follows that the more precise the determination of the indicator position is, the longer the process is, that is, the greater T is. Yet, according to general relativity, a clock which undergoes a position variation Δq, parallel to the direction of the field, undergoes also a variation of ΔT, proportional to T and Δq, of the time which it indicates. Therefore, the precision of the measurement of the photon energy, connected with the variation of the box mass, increases proportionally to the ΔT within which the photon goes out of the box. This means that, taking into consideration the gravitational field – something that Einstein did not do – the Heisenberg relation is still valid.

25. Paradoxically enough, the difference between rhetorical thought experiments and exploratory/clarifying thought experiments is so great that often when the first class is studied the second is neglected, and vice versa.
26. This is the version proposed by Misner et al. (1973, p. 187),
27. This is one of the principal reasons why some authors, such as Koyré (1960), confuse models and exploratory/clarifying thought experiments.
28. I am discussing the case with discrete time. If continuous time was considered, the model should be modified accordingly.
29. Sorensen (1992) devotes some very effective pages (pp. 218ff.) to clarifying what thought experiments are not. For example, they are not imagined experiments, fictional experiments (here in the sense that they are not the experiments proposed inside a novel or a movie), mythical experiments, simulation experiments and so on
30. Of course, the saturation has not to be intended à la Frege.
31. Note that there might be an objection of this kind: 'In this approach the circularity between laws or rules (*R*), that is, the elements of the nomological part, and counterfactuals, is not solved.' This is not true, as it can be understood by simply reflecting upon the fact that Weber's, Vaihinger's, and my own proposal belong to the Kantian tradition in which the nomological part 'comes before' (in a constitutive sense) what it covers.
32. Of course, there is the possibility that the rival hermeneutical regions can be more than two.

Bibliography

Adickes, E., *Kant und die Als-Ob-Philosophie* (Stuttgart: Frommans, 1927).

Allison, H.E., *Kant's Transcendental Idealism* (New Haven: Yale University Press, 1983).

Allison, H.E. 'Causality and Causal Laws in Kant. A Critique of Michael Friedman', in P. Parrini (ed.), *Kant and Contemporary Epistemology* (Dordrecht: Kluwer Academic Press, 1994), pp. 291–307.

Apelt, O., 'Die Kategorienlehre des Aristotles', *Beiträge zur Geschichte der griechischen Philosophie* (1891), 101–216.

Aristotle, *Analytica posteriora*, Engl. trans. *Posterior Analytics*, in *The Complete Works of Aristotle*, ed. J. Barnes (Princeton: Princeton University Press, 1984), Vol. 1, pp. 114–66.

Aristotle, *Analytica priora*, Engl. trans. *Prior Analytics*, in *The Complete Works of Aristotle*, ed. J. Barnes (Princeton: Princeton University Press, 1984), Vol. 1, pp. 39–113.

Aristotle, *Categoriae*, Engl. trans. *Categories*, in *The Complete Works of Aristotle*, ed. J. Barnes (Princeton: Princeton University Press, 1984), Vol. 1, pp. 3–24.

Aristotle, *De Interpretatione*, Engl. trans. *De Interpretatione*, in *The Complete Works of Aristotle*, ed. J. Barnes (Princeton: Princeton University Press, 1984), Vol. 1, pp. 35–8.

Aristotle, *Metaphysica*, Engl. trans. *Metaphysics*, in *The Complete Works of Aristotle*, ed. J. Barnes (Princeton: Princeton University Press, 1984), Vol. 2, pp. 1552–728.

Aristotle, *Physica*, Engl. trans. *Physics*, in *The Complete Works of Aristotle*, ed. J. Barnes (Princeton: Princeton University Press, 1984), Vol. 1, pp. 315–446.

Aristotle, *Topica*, Engl. trans. *Topics*, in *The Complete Works of Aristotle*, ed. J. Barnes (Princeton: Princeton University Press, 1984), Vol. I, pp. 167–277.

Armstrong, D.M., *Theory of Universals* (Cambridge: Cambridge University Press, 1978).

Armstrong, D.M., *What is a Law of Nature?* (Cambridge: Cambridge University Press, 1983).

Aspect, A., P. Grangier and G. Roger, 'Experimental Realisation of Einstein-Podolsky-Rosen-Bohm *Gedankenexperiment*: a New Violation of Bell's Inequalities', *Physics Review Letters*, 48(1982), 91–4.

Barker, P., 'Hertz and Wittgenstein', *Studies in History and Philosophy of Science*, 11(1980), 243–56.

Bayne, S.M., 'Object of Representation and Kant's Second Analogy', *Journal of the History of Philosophy*, 32 (1994), 381–410.

Bayne, S.M., 'Kant's Answer to Hume: How Kant Should Have Tried to Stand Hume's Copy Thesis on its Head', *British Journal for the History of Philosophy*, 8(2000), 207–24.

Beatty, J., 'The Evolutionary Contingency Thesis', in G. Wolters and J. Lennox (eds), *Concepts, Theories, and Rationality in the Biological Sciences* (Pittsburgh: University of Pittsburgh Press, 1995), pp. 45–81.

Beck, L.W., 'Is There a Non Sequitur in Kant's Proof of the Causal Principle?', *Kant Studien*, 67(1976), 385–9.
Beck, L.W., 'A Prussian Hume and a Scottish Kant', in L.W. Beck (ed.), *Essays on Kant and Hume*, (New Haven: Yale University Press, 1978), pp. 111–29.
Berkeley, G., *An Essay Towards a New Theory of Vision* [1709], in *The Works of George Berkeley*, ed. A.C. Fraser (Bristol: Thoemmes Press, 1998), Vol. I.
Berkeley, G., *Three Dialogues between Hylas and Philonous* [1713], in *The Works of George Berkeley*, ed. A.C. Fraser (Bristol: Thoemmes Press, 1998), Vol. I.
Black, M., 'Frege on Functions' [1954], in E.D. Klemke (ed.), *Essays on Frege* (Urbana: University of Illinois Press, 1968), pp. 223–48.
Boetius, A.M.T.S., *In Isagogen Porphyrii Commenta* (New York: Johnson Reprint Corporation, 1966).
Bohm, D., *Quantum Theory* (New York: Prentice-Hall, 1951).
Bohr, N., 'Discussion with Einstein on Epistemological Problems in Atomic Physics', in P.A. Schilpp (ed.), *Albert Einstein: Philosopher-Scientist* (Evanston: The Library of Living Philosophers, 1949), pp. 200–41.
Boltzmann, L., 'Über die Entwicklung der Methoden der theoretischen Physik in neueren Zeit' [1899], in L. Boltzmann, *Populäre Schriften* [1905], Engl. trans. in L. Boltzmann, *Theoretical Physics and Philosphical Problems* (Dordrecht: Reidel, 1974), pp. 1–198.
Boltzmann, L., 'Models', *Encyclopaedia Britannica* [1902], in L. Boltzmann, *Theoretical Physics and Philosphical Problems* (Dordrecht: Reidel 1974), pp. 211–20.
Boniolo, G., 'Kant's Explication and Carnap's Explication: the *redde rationem*', *International Philosophical Quarterly*, 43(2003), 289–98.
Boniolo, G., 'I modelli del mondo e il caso Galilei. Fra risultati empirici e questioni logico-epistemologiche', *Atti e Memorie dell' Accademia Galileiana di Scienze, Lettere ed Arti* (Padova-Firenze: Olschki, 2004), pp. 23–41.
Boniolo, G., 'A Contextualised Approach to Biological Explanation', *Philosophy*, 80(2005), 219–47.
Boniolo, G. and P. Budinich, 'The Role of Mathematics in Physical Sciences and Dirac's Methodological Revolution', in G. Boniolo, P. Budinich and M. Trobok (eds), *Mathematics and Physics: an Interdisciplinary Approach* (Dordrecht: Springer, 2005), pp. 75–96.
Boniolo, G. and S. Valentini, 'Vagueness, Kant, and Topology. A Study of Formal Epistemology', *Journal of Philosophical Logic* (2007, forthcoming).
Boniolo, G. and P. Vidali, *Filosofia della scienza* (Milano: Bruno Mondadori, 1999).
Boniolo, G., C. Petrovich and P. Pisent, 'Notes on the Philosophical Status of Nuclear Physics', *Foundations of Science*, 7 (2002), 425–52.
Brown, J.R., *The Laboratory of the Mind: Thought Experiments in the Natural Sciences* (London: Routledge, 1991).
Brown, J.R., *Smoke and Mirrors: How Science Reflects Reality* (London: Routledge, 1994).
Buchdahl, G., 'Causality, Causal Laws and Scientific Theory in the Philosophy of Kant', *British Journal for the Philosophy of Science*, 16(1965), 187–208.
Buchdahl, G., *Metaphysics and the Philosophy of Science* (Cambridge, MA: MIT Press, 1969).

Buchdahl, G., 'The Conception of Lawlikeness in Kant's Philosophy of Science', *Synthese*, 23(1971), 24–46; in L.W. Beck (ed.), *Kant's Theory of Knowledge* (Dordrecht: Reidel, 1974), pp. 128–50.

Burkamp, W., *Begriff und Beziehung. Studien zur Grundlegung der Logik* (Leipzig: Felix Meiner Verlag, 1927).

Burks, A.W., 'The Logic of Causal Propositions', *Mind*, 60(1951), 363–82.

Butts, R.E., 'Kant's Schemata as Semantical Rules', in L.W. Beck (ed.), *Kant's Studies Today* (La Salle: Open Court, 1969), pp. 290–300.

Butts, R.E., 'On Buchdahl's and Palter's Papers', *Synthese*, 23(1971), 63–74.

Butts, R.E., 'Induction as Unification: Kant, Whewell and Recent Developments', in P. Parrini (ed.), *Kant and Contemporary Epistemology* (Dordrecht: Kluwer, 1994), pp. 272–89.

Carnap, R., 'Testability and Meaning', *Journal of the Philosophy of Science*, 3 and 4 (1936–1937), 419–51 and 1–40.

Carnap, R., *Logical Foundations of Probability* (Chicago: Chicago University Press, 1950).

Cartwright, N., *How the Laws of Physics Lie* (Oxford: Oxford University Press, 1983).

Cassirer, E., *Substanzbegriff und Funktionsbegriff* [1910], Engl. trans. *Substance and Function* (Chicago: The Open Court, 1923).

Cassirer, E., *Kant Leben und Lehre* [1918], Engl. trans. *Kant's Life and Thought* (New Haven: Yale University Press, 1981).

Cassirer, E., *Philosophie der symbolischen Formen* [1923–1929], Engl. trans. *The Philosophy of Symbolic Forms* (New Haven: Yale University Press, 1963–1964), Vols 1–3.

Cassirer, E., *An Essay on Man: an Introduction to the Philosophy of Human Culture* [1944] (Garden City: Doubleday, 1953).

Chisholm, R.M., 'The Contrary-to-Fact Conditional', *Mind*, 55(1946), 289–307.

Condillac, E. Bonnot de, *Traité des sensations* [1754], Engl. trans. *Treatise on the Sensations* (London: The Favil Press, 1930).

Dretske, F.I., 'Laws of Nature', *Philosophy of Science*, 44(1977), 248–68.

Duhem, P., *La théorie physique: son object et sa structure* [1914], Engl. trans. *The Aim and Structure of Physical Theory*, 2nd edn (Princeton: Princeton University Press, 1954).

Dummett, M., *Frege: Philosophy of Language* [1973] (London: Duckworth, 1981).

Durkheim, E., *Les règles de la méthode sociologique* [1895], Engl. trans. in E. Durkheim, *The Rules of Sociological Method and Selected Texts on Sociology and its Method* (London: Macmillan, 1982).

Einstein, A., 'Elektrodynamik bewegter Körper', *Annalen der Physik*, 17(1905), 891–921.

Einstein, A., 'Über den Einfluss der Schwerkraft auf die Ausbreitung des Lichtes', *Annalen der Physik*, 35(1911), 898–908.

Einstein, A., B. Podolsky and N. Rosen 'Can Quantum-Mechanical Description of Physical Reality be Considered Complete?', *Physical Review*, 47(1935), 777–80.

Feyerabend, P., *Against Method* (London: Verso, 1975).

Feynman, R., *The Character of the Physical Law* (Cambridge, MA: MIT Press, 1965).

Fine, A., 'Fictionalism', *Midwest Studies in Philosophy*, 18(1993), 1–18.

Fraassen, B.C. van, *The Scientific Image* (Oxford: Oxford University Press, 1980).

Fraassen, B.C. van, 'Armstrong on Laws and Probabilities', *Australasian Journal of Philosophy*, 65(1987), 243–60.

Fraassen, B.C. van, *Laws and Symmetry* (Oxford: Oxford University Press, 1989).

Fraassen, B.C. van, *The Empirical Stance* (New Haven: Yale University Press, 2002).

Frege, G., *Die Grundlagen der Arithmetik. Eine Logisch-Matematische Untersuchung über den Begriff der Zahl* [1884], Engl. trans. *The Foundations of Arithmetic: a Logico-Mathematical Enquiry into the Concept of Number* (Evanston: Northwestern University Press, 1980).

Frege, G., *Begriffschrift* [1879], partial Engl. trans. in P. Geach and M. Black (eds), *Translations from the Philosophical Writings of Gottlob Frege* (Oxford: Blackwell, 1993), pp. 1–20.

Frege, G., 'Booles rechnende Logik und die Begriffsschrift' [1880–1881], in G. Frege, *Nachgelassen Schriften und wissenschaftlicher Briefwechsel*, ed. H. Hermes, F. Kambartel and F. Kaulbach [1969], Engl. trans. 'Boole's Logical Calculus and the Concept-script', in G. Frege, *Posthumous Writings* (Oxford: Basil Blackwell, 1979), pp. 9–46.

Frege, G., 'Funktion und Begriff' [1891], Engl. trans. 'Function and Concept', in P. Geach and M. Black (eds), *Translations from the Philosophical Writings of Gottlob Frege* (Oxford: Blackwell, 1993), pp. 21–41.

Frege, G., 'Über Sinn und Bedeutung' [1892a], Engl. trans. 'On Sense and Meaning', in P. Geach and M. Black (eds), *Translations from the Philosophical Writings of Gottlob Frege* (Oxford: Blackwell, 1993), pp. 56–78.

Frege, G., 'Über Begriff und Gegenstand', [1892b], Engl. trans. 'On Concept and Object', in P. Geach and M. Black (eds), *Translations from the Philosophical Writings of Gottlob Frege* (Oxford: Blackwell, 1993), pp. 42–55.

Frege, G., 'Ausführungen über Sinn und Bedeutung', [1892–95], in G. Frege, *Nachgelassen Schriften und wissenschaftlicher Briefwechsel*, ed. H. Hermes, F. Kambartel and F. Kaulbach [1969], Engl. trans. 'Comments on Sense and Meaning', in G. Frege, *Posthumous Writings* (Oxford: Basil Blackwell, 1979), pp. 118–25.

Frege, G., 'Kritische Beleuchtung einiger Punkte in Schröders *Vorlesungen über die Algebra der Logik*' [1895], Engl. trans. 'A Critical Elucidation of some Points in Schröder's *Vorlesungen über die Algebra der Logik*', in P. Geach and M. Black (eds), *Translations from the Philosophical Writings of Gottlob Frege* (Oxford: Blackwell, 1993), pp. 86–106.

Frege, G., 'Was ist eine Funktion?' [1904], Engl. trans. 'What is a Function?', in P. Geach and M. Black (eds), *Translations from the Philosophical Writings of Gottlob Frege* (Oxford: Blackwell, 1993), pp. 107–16.

Frege, G., 'Der Gedanke' [1918–19], Engl. trans. 'Thoughts', in G. Frege, *Logical Investigation*, ed. P.T. Geach (Oxford: Basil Blackwell, 1977), pp. 1–30.

Frege, G., 'Gedankengefüge' [1923–26], Engl. trans. 'Compound Thoughts', in G. Frege, *Logical Investigation*, ed. P.T. Geach (Oxford: Basil Blackwell, 1977), pp. 55–77.

Frege, G. 'Erkenntnisquellen der Mathematik und der mathematischen Naturwissenschaften' [1924–1925], in G. Frege, *Nachgelassene Schriften und wissenschaftlicher Briefwechsel*, ed. H. Hermes, F. Kambartel and F. Kaulbach [1969], Engl. trans. 'Sources of Knowledge of Mathematics and the Mathematical Natural Sciences', in G. Frege, *Posthumous Writings* (Oxford: Basil Blackwell, 1979), pp. 267–74.

Friedman, M., 'Causal Laws and the Foundations of Natural Science', in P. Guyer (ed.), *The Cambridge Companion to Kant* (Cambridge: Cambridge University Press, 1992a), pp. 167–97.

Friedman, M., *Kant and the Exact Sciences* (Cambridge, MA: Harvard University Press, 1992b).

Friedman, M., 'Kant and the Twentieth Century', in P. Parrini (ed.), *Kant and Contemporary Epistemology* (Dordrecht: Kluwer Academic Press, 1994), pp. 27–46.

Gabriel, G., 'Frege als Neukantianer', *Kantstudien*, 77(1986), 84–101.

Gale, R.M., 'On Some Pernicious Thought-Experiments', in T. Horowitz and G.J. Massey (eds), *Thought Experiments in Sciences and Philosophy* (Savage: Rowman & Littlefield Publishers, Inc., 1991), pp. 297–303.

Galilei, G., *Dialogo sopra i due massimi sistemi del mondo* [1632], in G. Galilei, *Le opere* (Firenze: Giunti Barbera, 1968), Vol. VII; Engl. trans. G. Galilei, *Dialogue Concerning the Two Chief World Systems – Ptolemaic and Copernican*, ed. S. Drake (Berkeley: University of California Press, 1953).

Gonella, F., 'Time Machine, Self-Consistency and the Foundations of Quantum Mechanics', *Foundations of Physics Letters*, 7(1994), 161–6.

Gooding, D., *Experiment and the Making of Meaning. Human Agency in Scientific Observation and Experiment* (Dordrecht: Kluwer, 1990).

Gooding, D., 'What is *Experimental* about Thought Experiment?', in D. Hull, M. Forbes and K. Okruhlik (eds), *PSA 1992* (East Lansing: Philosophy of Science Association, 1993), Vol. 2, pp. 280–90.

Goodman, N., 'The Problem of Counterfactual Conditionals', *Journal of Philosophy*, XLIV(1947), 113–28; in N. Goodman, *Fact, Fiction, and Forecast*, 3rd edn (Indianapolis: The Bobbs-Merrill Company, Inc., 1954–73), Ch. I.

Goodman, N., *Fact, Fiction, and Forecast*, 3rd edn (Indianapolis: The Bobbs-Merrill Company, Inc., 1954–73).

Guyer, P., 'Kant's Conception of Empirical Law', *Proceedings of the Aristotelian Society, Supplementary Volume*, 64(1990a), 221–42.

Guyer, P., 'Reason and Reflective Judgment: Kant on the Significance of Systematicity', *Nôus*, 24(1990b), 17–43.

Haken, H., *Synergetics. An Introduction* (Berlin: Springer Verlag, 1978).

Haldane, J.B.S., 'The Origin of Life', *The Rationalist Annual*, 3(1929), 148–53.

Helmholtz, H., 'Die neuren Fortschritte in der Theorie des Sehens' [1868], in H. Helmholtz, *Vorträge und Reden* (Braunschweig, 1896), pp. 265–366.

Henahan, S., 'From Primordial Soup to the Prebiotic Beach. An Interview with Exobiology Pioneer, Dr. Stanley L. Miller' (1996), www.accessexcellence.org /WN/NM/Miller.htm.

Hertz, H., *Die Prinzipien der Mechanik* [1894], Engl. trans. *The Principles Of Mechanics* (New York: Dover, 1956).

Hesse, M., *Models and Analogies in Science* (Notre Dame: University of Notre Dame Press, 1966).

Hume, D., *An Enquiry Concerning Human Understanding* [1748], in D. Hume, *The Philosophical Works* (Darmstadt: Scientia Verlag Aalen, 1964), Vol. 4, pp. 3–135.

Israel, G., *La visione matematica della realtà* (Roma-Bari: Laterza, 2003).

Jobe, E.K., 'Reichenbach's Theory of Nomological Statement', *Synthese*, 35(1977), 231–54.

Kant, I., *Kritik der reinen Vernunft* [1781–87, 2nd edn], Engl. trans. *Immanuel Kant's Critique of Pure Reason*, ed. N. Kemp Smith (London: Macmillan Education, 1989).

Kant, I., *Prolegomena zur einer jeden künftigen Metaphysik* [1783], Engl. trans. *Prolegomena to any Future Metaphysics*, ed. G. Hatfield (Cambridge: Cambridge University Press, 1997).

Kant, I., *Metaphysische Anfangsgründe der Naturwissenschaft* [1786], Engl. trans. *Metaphysical Foundations of Natural Science*, ed. M. Friedman (Cambridge: Cambridge University Press, 2004).

Kant, I., *Kritik der Urteilskraft* [1790], Engl. trans. *Critique of Judgement*, ed. W.S. Pluhar (Indianapolis: Hackett Publishing Company, 1987).

Kant, I., *Erste Einleitung* to the *Kritik der Urteilskraft*, Engl. trans. *First Introduction*, in *Critique of Judgement*, ed. W.S. Pluhar (Indianapolis: Hackett Publishing Company, 1987), pp. 385–441.

Kant, I. *Logik* [1800], Engl. trans. *Logic*, ed. R.S. Hartman and W. Schwarz (New York: Dover, 1974).

Kant, I., *Werkausgabe*, ed. W. von Weischedel (Frankfurt a.M.: Suhrkamp, 1956–1964), Vols 1–12.

Kant, I., *Briefwechsel* (Hamburg: Felix Meiner Verlag, 1986).

Kemp Smith, N., *A Commentary to Kant's 'Critique of Pure Reason'* [1923] (London: Humanities Press, 1950).

Kitcher, P., 'The Unity of Science and the Unity of Nature', in P. Parrini (ed.), *Kant and Contemporary Epistemology* (Dordrecht: Kluwer Academic Press, 1994), pp. 253–72.

Kitcher, P., 'Projecting the Order of Nature', in Patricia Kitcher (ed.), *Kant's* Critique of Pure Reason. *Critical Essays* (Lanham: Rowan & Littlefield, 1998), pp. 219–38.

Klemke, E.D. (ed.), *Essays on Frege* (Urbana: University of Illinois Press, 1968).

Kneale, W.C., 'Natural Laws and Contrary-to-Fact Conditionals', *Analysis*, 10(1950), 121–5.

Kneale, W.C., 'Universality and Necessity', *British Journal for the Philosophy of Science*, 12(1961), 89–102.

Koyré, A., 'Le de motu gravium de Galilée. De l'expérience imaginaire et de son abus', [1960], in A. Koyré, *Etudes d'histoire de la pensée scientifique* (Paris: Gallimard, 1981).

Krieger, L., 'Kant and the Crisis of Natural Law', *Journal of the History of Ideas*, 26(1965), 191–210.

Kripke, S., 'Semantical Analysis of Modal Logics', *Zeitschrift für mathematische Logik und Grundlagen der Mathematik*, 9(1963), 67–96.

Kuhn, T.S., 'A Function for Thought Experiment' [1964], now in T. Kuhn, *The Essential Tension* (Chicago: University of Chicago Press, 1977), pp. 240–65.

La Mettrie, J.O., de *L'histoire naturelle de l'âme* [1745], in J.O. de La Mettrie, *Œuvres philosophiques* (Paris: Coda, 2004).

Landau, L.D. and E.M. Lifšitz, *Teorija polija* (Moscow: Nauka, 1967).

Lee, K.S., 'Kant on Empirical Concepts, Empirical Laws and Scientific Theories', *Kant Studien*, 72(1981), 398–414.

Lewis, C.I., 'Experience and Meaning', *Philosophical Review*, 43(1934); in H. Feigl and W. Sellars (eds), *Readings in Philosophical Analysis* (New York: Appleton-Century-Crofts, 1949), pp. 128–45.

Lewis, D., 'Causation', *Journal of Philosophy*, 70(1973a), 556–67; in E. Sosa (ed.), *Causation and Conditionals* (Oxford: Oxford University Press, 1975), pp. 180–91.

Lewis, D., *Counterfactuals* (Cambridge, MA: Harvard University Press, 1973b).

Lewis, D., 'New York for a Theory of Universals', *Australasian Journal of Philosophy*, 41(1983), 343–77.

Lorentz, H.A., *The Theory of Electron* (Leipzig: Teuber, 1909).

Lotka, A.I., *Elements of Physical Biology* (Baltimore: Williams & Wilkins, 1925), new edn, *Elements of Mathematical Biology* (New York: Dover, 1956).

Lovejoy, A., 'On Kant's Reply to Hume', *Archiv für Geschichte der Philosophie* (1906), 380–407; now in M.S. Gram (ed.), *Kant: Disputed Questions* (Chicago: Chicago University Press, 1967), pp. 284–308.

Mach, E., *Die Mechanik in ihrer Entwicklung historisch-kritisch dargestellt* [1883], Engl. trans. *The Science of Mechanics. A Critical and Historical Account of its Development* (La Salle: Open Court, 1960).

Mach, E., 'Über Gedankeexperimente', *Zeitschrift für den physikalische und chemischen Unterricht*, 10(1896–1897), 1–5.

Mach, E., *Erkenntnis und Irrtum* [1905], Engl. trans. *Knowledge and Error* (Dordrecht: Reidel, 1975).

Massey, G.J., 'Backdoor Analyticity', in T. Horowitz and G.J. Massey (eds), *Thought Experiments in Sciences and Philosophy* (Savage: Rowman & Littlefield Publishers, Inc., 1991), pp. 285–96.

Matson, W.I., '"All Swans are White or Black". Does this Refer to Possible Swans on Canals on Mars?', *Analysis*, 18(1958), 97–8.

Maxwell, J.C., *Treatise on Electricity and Magnetism* [1873] (New York: Dover, 1952).

McCall, S., 'Time and the Physical Modalities', *The Monist*, 53(1984), 463–77.

Meinong, A., *Über die Stellung der Gegenstandtheorie im System der Wissenschaften* [1907], in R. Haller, R. Kindinger and R. Chisholm (eds), *Alexious Meinong Gesamausgabe* (Graz: Akademische Druk-u. Verlagsanstalt, 1973), Vol. 5.

Meyer, E., *Zur Theorie und Methodik der Geschichte* (Halle, 1902).

Miller, S.L., 'A Production of Amino Acids under Possible Primitive Earth Conditions', *Science*, 117(1953), 528–9.

Misner, C.W., K.S. Thorne and J.A. Wheeler, *Gravitation* (San Francisco: W.H. Freeman and Company, 1973).

Molnar, G., 'Kneale's Argument Revisited', *Philosophical Review*, 69(1969), 79–89.

Morrison, M. and M. Morgan (eds), *Models as Mediators* (Cambridge: Cambridge University Press, 1999).

Morton, A., 'Mathematical Models: Questions of Trustworthiness', *British Journal for the Philosophy of Science*, 44(1993), 659–74.

Nagel, E., 'Verifiability, Truth and Verification', *Journal of Philosophy*, 31(1934), 141–8.

Nagel, E., *The Structure of Science. Problems in the Logic of Scientific Explanation* (New York: Harcourt, Brace & World, Inc., 1961).

Newton, I., *Philosophiae Naturalis Principia Mathematica* [1687], Engl. trans. *The Mathematical Principles of Natural Philosophy*, introduced by I.B. Cohen (London: Dawson, 1968).

Norton, J., 'Thought-Experiments in Einstein's Work', in T. Horowitz and G.J. Massey (eds), *Thought Experiments in Sciences and Philosophy* (Savage: Rowman & Littlefield Publishers, Inc., 1991), pp. 129–48.

Norton, J., 'Are Thought Experiments Just What You Thought?', *Canadian Journal of Philosophy*, 26(1996), 333–66.

O'Shea, J.R., 'The Needs of Understanding: Kant on Empirical Laws and Regulative Ideals', *International Journal of Philosophical Studies*, 5(1997), 216–54.

Oparin, A.I., *Proiskhozhdenie Zhizni* [1924], Engl. trans. *The Origin of Life* (New York: Macmillan, 1938).

Pap, A., *An Introduction to the Philosophy of Science* (New York: Free Press, 1962).

Pargetter, R., 'Laws and Modal Realism', *Philosophical Studies*, 46(1984), 335–47.

Paton, H.J., *Kant's Metaphysics of Experience* [1936] (London: George Allen & Unwin, 1961), Vols 1–2.

Pearson, K., *The Grammar of Science* [1892] (London: J.M. Dent & Sons, 1943).

Poincaré, H.J., *La science et l'hypothèse* [1902–07, 2nd edn], Engl. trans. *Science and Hypothesis* (New York: Dover, 1952).

Poincaré, H.J., 'La fin de la matière', *Athenaeum*, 201–2(1906); in H.J. Poincaré, *La science et l'hypothèse* [1902–1907], Engl. trans. *Science and Hypothesis*, 2nd edn (New York: Dover, 1952), Ch. 14.

Popper, K.R., *Logik der Forschung* [1934], Engl. trans. *The Logic of Scientific Discovery* (London: Hutchinson, 1959).

Popper, K.R., *The Open Society and its Enemies* (London: Routledge and Sons, 1945), Vols 1–2.

Popper, K.R., *The Poverty of Historicism* (London: Routledge and Kegan Paul, 1957).

Popper, K.R., *Objective Knowledge. An Evolutionary Approach* (Oxford: Clarendon Press, 1972).

Porphyry, *In Aristotelis Categorias. Expositio per Interrogationem et Responsionem*, in *Porphyrii Isagogen et In Aristotelis Categorias Commentarium*, ed. A. Busse (Berlini: Academiae Litterarum Regiae Borussicae, 1887).

Prigogine, I. and I. Stengers, *La nouvelle alliance* [1979], Engl. trans. *Order out of Chaos* (New York: Bantan Books, 1984).

Quine, W.v.O., 'On What There is' [1948]; in W.v.O. Quine, *From a Logical Point of View* (Cambridge, MA: Harvard University Press, 1980), pp. 1–19

Ragnisco, P., *Storia critica delle categorie dai primordj della filosofia sino a Hegel* (Firenze: M. Cellini e C., 1871), Vols 1–2.

Ramus, P., *The Logike of the Moste Excellent Philosopher P. Ramus Martyr* [1574], Engl. trans. R. MacIlmain, contemporary edition C.M. Dunn (Northridge: San Fernando Valley State College, 1969).

Redhead, M., 'Symmetry in Intertheory Relations', *Synthese*, 32(1975), 77–112.

Redhead, M., 'Models in Physics', *British Journal for the Philosophy of Science*, 31(1980), 145–63.

Reichenbach, H., *Elements of Symbolic Logic* (New York: The Macmillan Company, 1947).

Reichenbach, H., *Nomological Statements and Admissible Operations* (Amsterdam: North-Holland Publishing Co., 1954).

Rescher, N., *Scientific Explanation* (New York: Free Press, 1970).

Rivadulla, A., *Ēxito, razón y cambio en física* (Madrid: Editorial Trotta, 2004).

Ross, W.D., *Introduction to Aristotle's* Metaphysics [1924] (Oxford: Oxford University Press, 1953).

Ruse, M., 'Are there Laws in Biology?', *Australasian Journal of Philosophy*, 48(1970), 234–46.

Russell, B., *Principles of Mathematics* (Cambridge: Cambridge University Press, 1903).

Russell, B. and A.N. Whitehead, *Principia Mathematica* (Cambridge: Cambridge University Press, 1910, 1912, 1913), Vols 1–3.

Salmon, W.C., 'Laws, Modalities and Counterfactuals', *Synthese*, 35(1977), 191–229.

Scaravelli, L., *Scritti kantiani* (Firenze: La Nuova Italia, 1968).

Schlick, M., 'Die kausalität in der gegenwartigen Physik', *Die Naturwissenschaften*, 19(1931), 145–62, Engl. trans. 'Causality in Contemporary Physics', in M. Schlick, *Philosophical Papers*, ed. H.L. Mulder and B.F.B. van De Vede-Schlick (Dordrecht: Reidel, 1979), Vol. 2, pp. 176–209.

Schlick, M., 'Sind die Naturgesetze Konventionen?' [1935] (Paris: Actes du Congrés de philosophie scientifique, 1936), 4; Engl. trans. 'Are Natural Laws Conventions?', in M. Schlick, *Philosophical Papers*, ed. H.L. Mulder and B.F.B. van De Vede-Schlick (Dordrecht: Reidel, 1979), Vol. 2, pp. 437–45.

Schlick, M., 'Meaning and Verification', *Philosophical Review*, XLV(1936), 339–69; in M. Schlick, *Philosophical Papers*, ed. H.L. Mulder and B.F.B. van De Vede-Schlick (Dordrecht: Reidel, 1979), Vol. 2, pp. 456–81.

Schopenhauer, A., *Über die vierfache Wurzel des Satzes vom zureichenden Grunde* [1847], Engl. trans. *The Fourfold Root of the Principle of Sufficient Reason* (La Salle: Open Court, 1974).

Schopenhauer, A., *Die Welt als Wille und Vorstellung* [1859], Engl. trans. *The World as Will and Representation* (New York: Harper & Row, 1968).

Schröder, E., *Vorlesungen über die Algebra der Logik (exacte Logik)* (Leipzig: B.G. Teubner, 1890–95; repr. Bronx: Chelsea Pub. Co., 1966), Vols I–III.

Sellars, W., 'Concepts as Involving Laws and Inconceivable without Them', *Philosophy of Science*, 15(1948), 287–315.

Sluga, H.D., *Gottlob Frege* (London: Routledge, 1980).

Smart, J.J., *Philosophy and Scientific Realism* (London: Routledge and Kegan Paul, 1963).

Sorensen, R.A., *Thought Experiments* (Oxford: Oxford University Press, 1992).

Stalnaker, R.C., 'A Theory of Conditionals' [1968]; in E. Sosa (ed.), *Causation and Conditionals* (Oxford: Oxford University Press, 1975), pp. 165–79.

Stalnaker, R.C. and R.H. Thomason, 'A Semantic Analysis of Conditional Logic', *Theoria*, 36(1970), 23–42.

Strawson, P.F., *The Bounds of Sense: an Essay on Kant's 'Critique of Pure Reason'* [1966] (New York: Methuen & Co., 1982).

Suárez, M., 'Scientific Representation: Against Similarity and Isomorphism', *International Studies in the Philosophy of Science*, 17(2003), 225–43.

Suárez, M., 'An Inferential Conception of Scientific Representation', *Philosophy of Science*, 71(2004), 1–13.

Thöle, B., *Kant und das Problem der Gestzmässigkeit der Natur* (Berlin: de Gruyter, 1991).

Tooley, M., 'The Nature of Laws', *Canadian Journal of Philosophy*, 7(1977), 667–98.

Tougas, J., 'Hertz und Wittgenstein, zum historischen Hintergrund der *Tractatus*', *Conceptus*, 75(1996), 205–27.

Trendelenburg, A., *Geschichte der Kategorienlehre, Historisce Beiträge zur Philosophie* [1846] (Berlin: Georg Olms Verlagsbuchhandlung Hildescheim, 1963).

Turing, A.M., 'The Chemical Basis of Morphogenesis', *Philosophical Transactions of the Royal Society B*, 237(1952), 37–72; in P.T. Saunders (ed.), *Morphogenesis: Collected Works of A.M. Turing* (Amsterdam: North-Holland, 1992), Vol. 3.

Vaihinger, H., *Die Philosophie des 'als ob'* [1911], Engl. trans. *The Philosophy of 'As If'* (New York: Barnes and Noble, 1968).

Vallentyne, P., 'Explicating Lawhood', *Philosophy of Science*, 55(1988), 598–613.

Vleeschauwer, H.J., *De La déduction trascendentale dans l'oeuvre de Kant* (Antwerpen: 'S Gravenhage, 1934, 1936, 1937), Vols 1–3.

Vollrath, E., *Studien zur Kategorienlehre des Aristotles* (Ratingen: A. Henn Verlag, 1969).

Volterra, V., 'Sui tentativi di applicazione delle matematiche alle scienze biologiche e sociali', *Annuario della Regia Università di Roma* (1901–1902), 3–28.

Volterra, V., 'Variazioni e fluttuazioni del numero d'individui in specie animali conviventi', *Memoria della Regia Accademia dei Lincei*, 2(1926a), 31–113.

Volterra, V., 'Fluctuation in the Abundance of a Species Considered Mathematically', *Nature*, 118(1926b), 558–60.

Volterra, V., *Leçons sur la théorie mathématique de la lutte pour la vie* [1931] (Paris: Gabay, 1990).

Waters, C.K., 'Causal Regularities in the Biological World of Contingent Distributions', *Biology and Philosophy*, 13(1998), 5–36.

Weber, M. 'Die "Objectivitität" sozialwissenschaftlicher und sozialpolitischer Erkenntnis' [1904]; in M. Weber, *Gesammelte Aufsätze zur Wissenschaftslehre* (Tübingen: Mohr, 1922), Engl. trans. 'Objectivity in Social Science and Social Policy', in M. Weber, *On the Methodology of Social Sciences* (Glencoe: Free Press, 1949).

Weber, M., 'Kritische Studien auf dem Gebiet der kulturwissenschaftlichen Logik' [1906]; in M. Weber, *Gesammelte Aufsätze zur Wissenschaftslehre* (Tübingen: Mohr, 1922), Engl. trans. 'Critical Studies in the Logic of the Cultural Sciences', in M. Weber, *On the Methodology of Social Sciences* (Glencoe: Free Press, 1949).

Wells, R.S., 'On Frege's Ontology' [1951], in E.D. Klemke (ed.), *Essays on Frege* (Urbana: University of Illinois Press, 1968), pp. 3–41.

Wittgenstein, L., *Tractatus Logico-philosophicus*, ed. C.K. Ogden [1922] (London: Routledge, 1996).

Witt-Hansen, J., 'H.C. Öersted, Immanuel Kant, and the Thought Experiment', *Danish Yearbook of Philosophy*, 13(1976), 48–65.

Index